High Protein Diet

By Fredrick Kerr

www.sunvillagepublications.com

High Protein Diet
By Fredrick Kerr

Copyright © 2010

No part of this publication may be reproduced, stored in a retrieval system or transmitted in any form or by any means, electronic, mechanical, photocopying, recording or otherwise, without prior written permission from the publisher.

Disclaimer: Neither the author nor the publisher accepts any responsibility for any injury, damage, or unwanted or adverse circumstances or conditions arising from use or misuse of the material contained within this publication. While every effort has been made to ensure reliability and accuracy of the information within, the liability, negligence or otherwise, or from any use, misuse or abuse arising the operation of any methods, strategies, instructions or ideas contained in the material herein is the sole responsibility of the reader. It is recommended that you consult a physician before making any changes to your diet or health, or undergoing any new exercise regime.

Cover photo credits: © Cammeraydave/Dreamstime.com

www.sunvillagepublications.com

Cover design by www.WebCopyAlchemy.com

CONTENTS

INTRODUCTION by Leonid Kotkin, M.D. 7

ONE: DON'T COUNT CALORIES- BUT CALORIES DO COUNT! 9

Part One: DIET

TWO: WHAT IS A CALORIE? 17
THREE: REDUCE WITHOUT TEARS! 23
FOUR: WHAT IS A PROTEIN? 29
FIVE: A HIGH-PROTEIN DIET 35

Part Two: DO ITI

SIX: WHAT FOODS CAN YOU EAT? 41
SEVEN: COOKING TO SAVE NUTRITION 73
EIGHT: THE HIGH-PROTEIN MENU 81
NINE: SNACK AND STAY SLIM 112

Part Three: STICK TO IT!

TEN: FATS, FADS AND FALLACIES 121
ELEVEN: WHAT ABOUT EXERCISE? 138
TWELVE: WEIGHT WATCHING 145
THIRTEEN: DOCTOR ABOVE ALL 156

INTRODUCTION
by Leonid Kotkin, M.D.
(Author of *Eat, Think and Be Slender)*

This is a sensible book on an important subject. There are no "gimmicks" in it—it only offers good advice.

Books on how to lose weight have flooded the market for several decades. Many new ones appear every year. The trouble is that the book with a gimmick always sells well. This is because many people are, unfortunately, magical-thinking. They are always eager to attempt a strange or weird diet.

In fact, it would be interesting to do a study of the personalities of those obese people who eagerly try anything new, who are never discouraged and try again and again—just as long as it is something new; and the weirder the better.

For a while, thousands of Americans were drinking formulas, like infants. When this failed to reduce them permanently, they tried eating unlimited fat in order to lose fat. Safflower oil seed still sells by the ton in spite of the fact that the United States Public Health Service has a suit against a number of persons and a manufacturer because of false statements they have made that it is effective in losing weight.

And these Americans will constantly remain fair game for quacks and miracle-men who abound. They will constantly seek out "sure cure" treatments. These people will probably shun a book that offers a sensible approach to weight reduction, a program that is gimmick-less and offers no overnight miracles.

There is only one thing about this book that I cannot approve. The author, in his enthusiasm, has rather over-emphasized my reputation and importance. The reader should know that I read the book after it was a *fait accompli*—it was too late to change this flattering evaluation.

1: DON'T COUNT CALORIES—
BUT CALORIES DO COUNT!

"What, another diet book?"

You undoubtedly asked that question the moment you picked up this book.

The answer is: "Yes, but there is a reason!" There is a real need for a book on dieting to be used by people who are looking for a simple, honest and intelligent approach to the whole question of weight control.

Although it may be true that few people have *literally* eaten themselves to death, most American health authorities are quick to agree that a good many people in this country are eating their way to illness and, in many cases, to an earlier-than-necessary death.

Obesity—that is, overweight—may not be listed as the "cause of death," but there is no doubt that overweight has become one of the most pressing health problems in this country today. It is certainly one of the paradoxes of our age that many millions of Americans of all ages are suffering malnutrition from excessive intake of food, while in other areas of the world many more millions of people suffer from starvation.

The American who is overweight—and about one-third of our population is—may be a symbol of the opulence of the country. But this overweight American is also the unhappy evidence that we are not doing a very good job of learning how to build and how to maintain a physically fit populace.

So much emphasis has been put on the "fun" of eating that for many Americans the pursuit of happiness seems to be an endless stuffing of stomachs. The basic purpose of eating seems to have been forgotten, however. Its purpose is to provide the essential nutrients which bodies need for growth and maintenance. It should be as easy to learn healthful eating habits as it is to learn poor ones.

But since Americans have learned such poor eating habits, they must spend many difficult hours trying to unlearn these habits—habits which have contributed to their undesirable excess weight.

The overweight person who seriously desires to shed excess pounds will find that fad diets designed to take off weight fast and furiously seldom accomplish the goal of keeping the desired weight over the long pull. Too many diets today are based on extreme theories, too many diets promise miracles overnight. Some advocate easy methods, but don't work; others are so rigid they won't be followed.

There are no “miracle” diets. And while modern science is always moving rapidly and has a way of outdating knowledge quickly, nonetheless the basic concepts of good nutrition seem to be well established; and a diet that is conceived on these basic concepts is the surest way of achieving a more healthy existence.

There is nothing good that can be said for overweight. It is a barrier between you and health—between you and beauty. It is an enemy to efficiency and fun. It is a threat to long life.

Statistics show that the death rate of those who are chronically, but only moderately, fat is one and one-half times higher than that of those who keep themselves trim and sum. The death rate of the markedly obese zooms to one and three-fourths times that of people of normal weight. Therefore, staying slender is much more than a matter of good looks and fun. It is a duty that you owe to yourself.

Most people think of dieting to lose weight as a dreary enterprise—they think of it as something tiresome, something expensive, and something that will be a complete bore, not only to themselves but to their families and to their friends. Well, it need not be this. Dieting to grow slim need not be punishment, despite the "curse of the calorie," which Americans have had pounded into their fat-filled bodies for the last two or three decades.

In attempting to ease the burden of the extra weight they have carried, the more than fifty-five million people in this country who are overweight have become calcu-latingly conscious of the calorie. Hundreds of diets have been created to help the overweight person reduce his calorie intake. Others have been created to confuse the dieter with the fact that there is no necessity even to regard calories.

Actually, this book attempts to clarify once and for all that while it is wrong to say *"calories don't count,"* it is equally wrong to think that *"you must count calories."*

The concept upon which this book is based is that there is no real necessity for counting calories, if—and that is a big IF—you eat a properly conceived nutritional diet.

It is obvious that there are many ways to lose weight. Some are good—they keep you energetic and well as they help you lose the unwanted pounds. Some are bad—they damage the health and are, in the long run, useless.

This book has attempted to point out to you why so many of these fad diets are dangerous. In addition, it will show how you, by simple planning of your daily menu, you can assure yourself a steady and safe loss of weight.

If every one of you had to carry around your excess weight in a bag chained to your arm, like a bank messenger, you would very quickly find a way to get rid of this weight. Yet that inconvenience is a trifle compared to the harm that the unneeded fat you are now carrying on your body can do. In effect, by being overweight you are gambling with your life. Overweight is, in reality, the nation's Number One health hazard. And while medical discoveries in the past half-century or more have worked wonders in providing better health and longer life for all members of the family group, it is increasingly apparent that Americans are either unwilling or unable to do anything to combat this major health hazard.

The importance of proper weight to health is not being exaggerated. You undoubtedly appreciate that weighing too much can be a handicap in social and economic life. But you may find it more difficult to understand how serious a menace to health it is. Yet studies by life insurance companies, Army records, surveys conducted by health agencies, hospital records, and many private research projects show that:

1. There are at least fifty-five million—if not more-people in this country who are at least ten per cent overweight. For example, there are people who should weigh, say, 130 pounds but who weigh at least 143 pounds or more. Of course, these figures must be expressed as a percentage, rather than in pounds, or they would be meaningless.
2. Of these fifty-five million or more people, about half, some 28,000,000, are much more than ten per cent overweight. In fact, close to ten million of them are at least twenty per cent above normal.
3. The danger-line in respect to excess weight is about ten per cent. If you exceed your normal weight by this amount you should certainly reduce in order to stay healthy.
4. If you are in the ten- to twenty-per-cent-excess class, you have a real liability and you should not delay in taking off that weight.
5. If you are more than twenty per cent overweight your condition is extremely serious. Reducing will help you, but you may have some other problems of long standing that won't entirely be cleared up. It is most important that you consult a doctor, and with his help get down to normal weight as quickly as possible so that you will have a better chance to combat the other problems.

The high-protein diet prescribed in this book will give you a wide choice of easy menus for sure and steady and safe weight loss. In addition, it will point out diets that are dangerous, it will tell you how exercise can help you tone up your muscles—while not actually reducing much, if any, weight—and fill you in on all the other important aspects of dieting.

In the pages that follow, for instance, you will discover what calories really are. A whole chapter is given over to evaluating calorie concepts; the calorie is defined for you; and the difficulty of a calorie-count diet in any way, shape or form is pointed out.

The value and simplicity of the high-protein diet is then discussed. You learn how proteins are the building blocks of life. You will discover the theory of dieting on protein foods—a theory which has been successfully proved over many years—as opposed to diets that call for high quantities of fats and/or carbohydrates.

You will be given a list of foods that you can eat. You will find out what the high-protein foods really are and you will see what foods have to be avoided in order to lose weight.

You will discover how not to waste the vitamin and mineral values in cooking, in order to get as much nutrition out of the food you eat as possible.

You will be told how to plan a high-protein menu, and you are even given a number of recipes that will help to brighten your reducing program as well as show you how other foods can be adapted to the high-protein concept.

You will discover the trick of taking a snack and still staying slim. You have, perhaps, already found out how snacking has been your downfall in the past. Yet you will be shown how you can still sneak snacks between meals and stay on your diet.

As a dieter who has undoubtedly tried losing weight before, you will find out why the weird diets you have tried before have not worked, why they cannot work and why, even though you may have lost some weight, you could not keep that weight off.

You will be taught the all-important concept of weight watching.

Above all, you will recognize that there is no magic potion or formula that can make you thin. There are no miraculous pills or starvation regimens or other hocus-pocus about it. Every time an abracadabra plan is announced in the advertisements, just ask yourself what became of the last "reducing discovery" that recommended itself in the same glowing terms. Also, do the same with those quick-trick diets that guarantee to cut your weight in no time by abstaining from or accenting some one particular food.

Successful reducing means a permanent weight loss. It is a strictly personal problem, a problem that only you can lick. So, *diet, do it,* and *stick to it!*

Part One: DIET

2· WHAT IS A CALORIE?

As one of the some fifty-five million people in this country concerned about your weight because you are heavier than you should be, you are undoubtedly extra-calorie-sensitive.

You have been pounded on all sides with too many calories in this, too many calories in that—in fact, just too many calories.

You have been told by your doctor, you have read in newspapers and magazines, and even in many books, that unless you control your calorie intake, you are leaving yourself open for a wide variety of ailments that lead but to the grave.

The calorie-counting concept has been filled with the mumbo-jumbo of mystery. The countless factors involved in prescribing caloric intake and the eating routine for a reducing dieter would require you to have the knowledge and the experience of a physician who has been in practice for twenty years or more.

For example, if you were to try to unravel the mystery, you would have to take all of these factors into consideration:

Your sex Your age
Your weight
Your height
Your frame size
Your general health
The proportion of muscle to fat in your body

Your metabolism (that is, the rate at which you burn calories)
Your physical activity
Your previous eating habits

At first glance, if you were to begin to work out a formula involving all of these ten factors you could work from now until doomsday and never quite achieve a satisfactory diet that would bring you to a healthy weight.

For this reason, nutritionists and medical experts have worked out a variety of calorie-counting diets. There are 600-calorie diets, 900-calorie diets, 1200-calorie diets, 1500-calorie diets, 1600-calorie diets, 2000-calorie diets, 3000-calorie diets, and several dozen in-betweens that would drive a non-mathematically-minded person slightly dizzy.

Actually, there is nothing mysterious about a calorie. It is nothing more than a unit of measurement.

For example, we measure length by inches and feet. We measure heat by degrees on a thermometer.

Similarly, we measure the energy-producing quality of food by a unit called a calorie. The calorie is nothing more than a measurement of heat, or energy, produced by food as it burns in the body. It can also be measured as the food is burned in a vacuum in a laboratory, for this is the way in which caloric values are determined.

The body is just like a furnace. And as fuel burning in a furnace produces a measurable quantity of heat—which is a form of energy—so food being "burned" as it is utilized by the body uses a measurable amount of energy. The heat in a furnace is measured by British Thermal Units, or BTU's. The heat in the body is measured in calories.

Because foods are made up of different substances, a pound of one food will give off many more calories than a pound of another food. This is the same as the principle of a ton of coal burning longer and producing more heat than a ton of wood.

However, the comparison between the body being fed

by food and the furnace with fuel ends at this point. If you put too much fuel in a furnace, the heat goes up the chimney and radiates through the metal into the air, disappearing. It doesn't cling to the furnace making it fatter. When you eat more food than you need, however, the extra energy is not eliminated. The body stores it for future use by turning it into fat.

Normal body temperature is 98.6 degrees and the food we consume is burning, or oxidizing, in our bodies to replace the heat that we lose by physical exertion, by bodily radiation, by breathing, and other ways. Even while we sleep, we burn up the food in our bodies at the rate of about 1,700 calories a day for a body that weighs 160 pounds.

Specifically, a calorie is the amount of heat that is needed to raise the temperature of a pint of water four degrees Fahrenheit. Actually, one lone calorie will keep a sleeping body running for about fifty-one seconds.

To break it down into more readily understandable terms, consider that one gram of a food (which is less than one-quarter teaspoon) that is essentially protein or essentially carbohydrate will supply four calories of heat or energy. And one gram of fat will give nine calories. Of course, any calculation for any specific food has to be "approximate." Food is perishable and the number of calories for any given food depends upon its freshness, the time of year, the variety or type of food, its maturity, where it was grown, the amount of sun and water to which it was exposed, the conditions of the soil in which it was grown, and a host of other variable factors. And while it is impossible to be correct down to the single calorie, that amount of accuracy in calorie-counting is not necessary since variations will average out.

The body needs different amounts of energy for different occupations. The National Research Council, for instance, has recommended the following requirements:

	AGE	WEIGHT	HEIGHT	CALORIES
MEN	25	154	69 inches	3,200
	45	154	69 inches	3,000
	65	154	69 inches	2,550
WOMEN	25	128	64 inches	2,300
	45	128	64 inches	2,200
	65	128	64 inches	1,800
Pregnant (last 4½ months)				+300
Breast feeding (850 ml. daily)				+1,000
BOYS	13-15	108	64 inches	3,100
	16-19	139	69 inches	3,600
GIRLS	13-15	108	63 inches	2,600
	16-19	120	64 inches	2,400
CHILDREN	1-3	27	34 inches	1,300
	4-6	40	43 inches	1,700
	7-9	60	51 inches	2,100
	10-12	79	57 inches	2,500

These levels are intended to cover individual variations among most normal persons living in the United States under the usual environmental stresses. The calorie allowances apply to persons usually engaged in moderate physical activity. It must be realized that for office workers and any other persons in occupations that are essentially sedentary, these allowances are excessive. For example, the forty-five-year-old male in the above table will only require 2,400 calories a day if he has a sedentary job. And the same man, doing heavy labor, will require 4,500 calories a day.

So you see that adjustments have to be made depending upon what your build is, what kind of work it is you do essentially, and, of course, your age, sex and even environmental temperature. And if you want to lose weight, you have to cut down by a certain prescribed number of calories to begin to lose that weight.

Actually, when you see how many calorie-counting diets there are, it would seem as though cutting down on caloric intake is a wonderfully simple reducing plan in itself. The trouble is, however, that it is not.

A little knowledge of calories can be a dangerous thing. Caloric values of food have nothing to do with measurement of mineral or vitamin or protein content of food. And these are elements which we must have in order to keep alive.

Calories measure energy only—energy which is derived from food and drink. For example, a half-pound of steak contains exactly twice the number of calories as a quarter-pound. But, it is actually less fattening to eat a half-pound steak than a quarter-pound one. The reason for this is that if you eat more protein, you will eat less carbohydrate, and it is this conversion of carbohydrate into fat by the human body that is one of the main causes of obesity.

Unfortunately, humans share with pigs the ability to easily convert carbohydrates into fat. Such fat, forming around the heart and between the muscle fibers, causes disease. Protein, on the other hand, does not easily become fat. That part of carbohydrate which we consume and which is not converted into energy does, on the other hand, easily become converted and stored away in the body as fat.

For example, alcohol supplies fuel for energy in the quickest and most easily assimilated way. If you are supplied with sufficient energy from alcohol, you do not burn up as much of the carbohydrates and these subsequently turn into fat. That is why if alcohol is taken in large quantities, fat will eventually be produced in the body.

All foods can be fattening if we eat more of any one, or any combination, than the body needs—because all food, as has been pointed out, has caloric value. There are calories in bran and soft drinks, in alcohol, in lemon peel, in pickle juice, just as there are in cake and steak. It's not a matter of there being "bad calories" or "good calories" just as there are no "bad foods" or "good foods." All foods are good *when properly used.*

For example, a large glass of whole milk has about 165 calories. A big piece of fudge has about the same. What the calorie count does not show is that milk is packed with proteins, vitamins and minerals. Animal proteins, which are obtainable in milk, eggs, cheese and lean meat and fish, are one of the essentials to health. They are all-important in the replacement of worn tissue. A lack of sufficient vitamins and minerals will create illness.

On the other hand, fudge contains barely enough protein, vitamins and minerals to register. Being largely composed of sugar and fat, fudge provides little, though fast, energy. It does not make your skin glossy and clear. It does not make your eyes bright. It does not build bone or blood or lean tissue.

That is not to say that sugar and fat have to be despised. They have a useful place in a normal diet. But a reducing diet is not a normal diet. And if your doctor has figured out that you would need 1,650 calories a day in order to reduce, by taking ten pieces of fudge to make up the required calorie intake you are not going to put yourself on a healthy reducing diet. All you're going to do is put yourself in the hospital in short order.

And, of course, ten glasses of milk a day—with nothing else—would certainly lead to trouble too. Though if you were to live on milk alone, the trouble would not come about so rapidly, because milk supplies a great many of the food elements needed for health.

* * * * *

Rx *for Reducing:*

1. **All foods contain calories.**
2. **Calories are only the measure of energy in food.**
3. **Human energy requirements vary from per son to person, and are based on many factors.**
4. **Calorie-counting reducing diets require the genius of a mathematician to be workable.**

* * * * *

3: REDUCE WITHOUT TEARS!

From the point of view of good nutritive planning, the counting of calories has an important place. As you read in the previous chapter, the amount of food-intake you require to compensate for energy-output depends on your sex, age, weight, height, frame size, general health, physical activity, etc. This is the calorie measurement.

But, as an overweight person attempting to reduce, you can only run into a variety of troubles if you attempt to reduce on a calorie-count diet alone.

As has been previously pointed out, to create a reducing diet solely on a caloric factor, you would have to have the knowledge and experience of a physician. Therefore, planning such a diet on your own becomes highly impracticable and virtually unworkable. In fact, your own doctor, if he were to put you on such a diet, would much prefer to do it under conditions which medicine terms "clinical" By that is meant that you would remain under constant surveillance, preferably in a hospital, so that the proper foods could be served to you at the proper times, and so that you could not in any way "cheat" on your diet—since all food you receive would be strictly supervised.

A reducing diet is a waste of time unless it does two things besides take pounds off you:

1. It must keep you at peak energy.
2. It must keep you in good health.

Your body is made up of many different materials.

These materials all come from food and drink. But, you need many different foods each day because different foods supply different ingredients necessary to good health.

Some foods go into the upkeep and replacement of working tissue, such as muscles, and keep your blood red and healthy. These foods also supply energy. They are sometimes called the "protective" foods.

Other foods do little or nothing for the upkeep of the body. They are sometimes called the "empty" foods, because they are lower, or lacking, in the essential vitamins, minerals and proteins. They mainly supply energy and have a valuable place in the normal diet—again emphasizing that a reducing diet is not a normal diet.

When you diet to reduce, what you want to lose is fat and, with that fat, those extra inches around the waist. You do not, however, want to lose lean tissue. Loss of weight from lean tissue is dangerous, and useless.

So if you foolishly adopt a substandard diet for the sake of a quick loss of weight, you will get no permanent results. And if you go on a starvation diet, you are bound to lose some fat but you will also lose with it a considerable amount of lean tissue. Therefore, you will be punishing yourself by going on an overly drastic reducing diet.

A diet so drastic that it approaches the starvation level is bad for another reason too. It interferes with elimination which requires a certain amount of bulk. This bulk can be supplied by many different foods. During a reducing diet, this bulk should be supplied by food that also contains the greatest amounts of the vitamins, minerals and proteins you need for health—whether you are reducing or not.

The purpose of this book is, essentially, to show you the easiest, and the most nutritive, way to lose weight.

Up to now, it should be obvious to you that the concept of calorie-counting, while it is logical from the medical point of view, is one which can create for you a great amount of difficulty. For you to be a true calorie-counter, you are going to have to carry with you a chart of caloric values of all foods. These are available. They have been published in newspapers, magazines, by various organizations interested in your health, and even by commercial publishers who have tried to make it easy for you by preparing such calorie charts in small vest-pocket or purse-size pamphlets.

This means, therefore, that every meal must be carefully weighed beforehand. You must attempt to find out just how large a portion you are going to have of every food that comprises the meal, and work out the caloric content of the food based on the size of the portion.

This may not be too difficult if you are preparing your own meal at home—providing that you have very accurate means of weighing and measuring all your foods. It will prove far more difficult—next to impossible—when you are eating out, as many of us do from time to time depending upon what kind of work we do, where we live, etc.

In order to facilitate dieting, therefore, this book has been created around the concept that you need not count calories.

In fact, the admonition can be made right here:

DON'T COUNT CALORIES!

But, it isn't as simple as *just* not counting calories. You can go on a "don't count calories" diet providing you keep a watchful eye on what you eat. It's going to be useless to you if you start to diet on a "don't count calories" concept, if you also don't plan your food intake.

Good food intake, however, is something that will be discussed later. There are a few simple principles that can help you plan meals that are easy to make, delicious to eat, and above all, good for you while you are losing weight.

But before we get to that, it is necessary to re-emphasize the fact that you do not have to count calories. In fact, if you start counting calories you are going to run the great risk of failing in your plan to reduce your weight level.

The fact that you are reading this book means that you understand that you have a weight problem—whether you are only five pounds overweight (and unable to fit into last season's clothes) or fifty or more pounds overweight (thus leaving yourself open to other bodily deterioration).

The chances are better than a hundred to one that this is not the first diet book you have read. In fact, the chances are extremely good that you own several diet books, that you have bought issues of magazines that you would not ordinarily buy simply because the cover stated that there was an article on the inside concerning dieting, and that you have clipped various articles from newspapers that concern themselves with dieting and weight-reduction problems.

If this is true, and the odds are extremely high that it is, then why are you reading this—another book on dieting?

The answer is simple. You are now reading this book because you have not been successful over the long haul before.

That is not to say that you haven't lost weight successfully on other diets. You probably have. You may even have lost all the weight you wanted to, though the statistics available indicate that in most instances you have never completely successfully dieted down to the weight that you should be.

But, whether you dieted successfully—that is, to the desired weight—or not, the reason you are reading this book now is that you face the problem of dieting once again.

In other words, you have found that not only is overweight an illness, it appears to be a chronic illness. No sooner do you cure yourself than you are afflicted again. And, quite possibly, again and again and again!

Wherein, then, lies the problem? Is it that the "cure"

you undertook didn't work? The answer is obviously not an outright "no." You know that you have been able to lose weight on various diets that you have been on previously. In fact, you may have even lost all the weight you wanted to lose.

But a cure cannot be termed successful if it doesn't "stick." If the illness is going to become chronic, then the cures you have taken in the past have been useless in that they have only provided temporary relief. And a plan of temporary relief, if it has to be used again and again, is really a pointless plan when complete relief is possible.

The point being made here is a simple one. Calorie-counting can be an effective way of planning a diet to lose weight. The difficulty is, however, that a diet that comprises only calorie-counting is going to give you the kind of foods that are wrong for a reducing diet. A calorie-counting diet provides a certain leeway in dieting that, while pleasant, is destructive, because it is too lenient in the type of foods permitted. The diet advocated by this book will also provide you with leeway—in fact, far more leeway in many respects than you have been used to with calorie-counting diets. But the leeway provided by the diet in this book is one which adds to the health-giving aspects of a diet without adding to the destructive aspects of a calorie-counting diet.

To be a calorie-counter you have to be virtually the equivalent of a chartered accountant. You have to carry around with you a master chart, and a notebook to record your daily intake on a meal-by-meal and snack-by-snack basis. You end up spending almost as much time recording the number of calories taken in as you do in actually taking them in. And if anything is going to spoil the enjoyment of eating, it's going to be the tedious task of calorie-count collecting.

* * * * *

Rx *for Reducing:*

As a calorie-counter, you are going to run into the most detrimental aspects of dieting:

1. You must measure every portion of food you eat at every meal, and even between meals.
2. You must work out the caloric value of this specific portion.
3. You must record the total number of calories for each portion and maintain a daily level that is within the diet plan you have under taken.
4. The freedom of a calorie-count diet provides for the intake of foods which are essentially "empty" foods—foods lacking sufficient vitamins, minerals and proteins.

* * * * *

4: WHAT IS A PROTEIN?

Nine persons out of every ten who have ever been on a diet have asked, at one time or another, "What is a protein?"

To most persons who attempted to reduce, and to other overweight persons who have contemplated the idea but have never done anything about it, the principle of diet has always been calorie-counting. This, as has already been pointed out, is very scientific. The difficulty is that in order to follow a calorie diet to the finest degree you have to carry a complete chart of all foods, with their caloric values, and a system of weighing and measuring every portion you eat. In addition, you have to know exactly what went into the preparation of everything you consume.

It becomes obvious, then, that in order to maintain a calorie-counting diet, you are undertaking a task that is virtually impossible.

But diet means "proper food." And one of the basic "proper foods" that is essential to all life is something called protein. Protein is basically the "stuff of life."

Your body is made up largely of protein: your skin, muscles, internal organs, nails, hair, brain, and even the base of your bones. Only when protein of excellent quality is supplied can every cell in your body function normally and keep itself in constant repair. Since your muscles contain a greater amount of protein than the other parts of your body, all you have to do is look at yourself in the mirror to estimate roughly the adequacy of your protein intake.

If our mirror shows you an erect body, then you can suspect that you have strong, well-nourished muscles. For it is only such muscles that can automatically hold your body erect. When your muscles have not received the food necessary for their repair, they must lose their elasticity just as an old rubber band does. The result is poor posture.

Without conscious effort, a person who is healthy holds his head high and his chest out. His shoulders and abdomen are flat, his feet have well-defined arches, and his step is rhythmical.

Protein is the primary part, then, of all organs and tissues in all animals and plants. It is the only truly living portion of any tissue. Actually, protein is an extremely complex matter. In fact, it is so complex that by contrast starches, sugars and fats are infinitely simple in their structure. And while starches, sugars and fats serve the function of nourishing tissue or acting as reserve material in the body, if the body has an insufficient or inadequate amount of protein, there is an almost immediate breakdown of the body's natural functions and restorative powers that eventually make for ill health of many natures.

The word protein was derived from a Greek word meaning "first," since food contains five main constituents and since the first in order of importance is protein. Protein is "first" because it supplies the complex chemical molecules from which life itself is created. From it, new tissue is formed, thus permitting growth, and dead or used-up tissue is replaced. Scientific research has shown conclusively that without protein there could be no life as we know it.

Protein in different foods is made up of varying combinations of twenty-two simpler materials called amino acids. If need be, the body can make its own supply of more than half of these amino acids. But the remaining amino acids must come ready-made from food. And to get the best use of these special ones, the body needs them all together, either in one food or in some combination of foods.

Protein has other peculiar and remarkable properties. Our knowledge of these has not been reached by scientific research but has been discovered by experience. For instance, it has been discovered that people who habitually eat a high-protein diet—that is to say, one which contains relatively small amounts of fat and carbohydrates—have more endurance and energy. They are summer, though not necessarily lighter, than those who habitually eat more carbohydrate and less protein.

When protein *is* the predominant feature of our diet, we are more energetic, our endurance is greater, we tend to be slimmer, unnecessary and unsightly fat disappears, we are more resistant to infectious diseases, and we have a heightened sense of well-being.

There is more than one scientific theory as to why this is so. But they are too complicated to go into here. Such are the facts, however. Athletes, when in training, eat a high-protein diet. It was once thought—and this theory was held for a long time—that their increased powers of endurance and general fitness, as well as their absence of excess weight, were due to the amount of physical exercise that they took. It has, however, been proven that this is not the whole story, nor perhaps even the most important part of it.

Retired athletes, leading a sedentary life, if they continue on their high-protein and low-carbohydrate, low-fat diet, will in general continue to feel very fit and keep slim, while those who abandon this diet will tend to run to fat.

It appears, then, that the body must have adequate amounts of protein at all times. Where does one get this protein?

Meat, fish, eggs and cheese contain large amounts of protein. And these rich dietary sources of high-quality protein should be included in your daily diet in order to maintain health and youthfulness.

Since your body structure is largely protein, an undersupply can bring about aging with depressing speed. But when proteins are eaten, the body's digestive process breaks them down and passes them into the blood and carries them throughout the body. The various cells of the body select those amino acids they need and use them to construct new body tissue as well as antibodies, hormones, enzymes and blood cells. During every instant of life, body proteins are being broken down by enzymes in your cells, and if your health is to be maintained, amino acids must be available for immediate replacement.

Of course, you will say, proteins include calories. Everything includes calories, but proteins proportionately have the lowest calorie content of the three basic food types—fats, carbohydrates, and proteins. Therefore, without having to work mathematical miracles and without having to juggle complicated figures, you can eat the foods on a protein diet which can supply only the lowest possible caloric content. Without a superabundance of calories, and with the food you eat being pure tissue builder, your body has the least chance to convert into fat any excess calories from your meals.

For the energy you need for all your daily tasks, therefore, your body must utilize the fat that has been stored away in your tissue previously. The result is that the stored energy is used instead of fresh energy being adapted, and your excess pounds are worn away, steadily and surely.

To get a better picture of what happens to protein when it enters the body, let us look at the basic process of digestion and metabolism in the body. In digestion, the food is broken down into the component substances. In metabolism, these substances are rebuilt into living matter.

Basically, what happens in digestion is that substances which the body manufactures—enzymes—attack the food that you have eaten. These enzymes bring about a chemical change in that food and break it down into simpler and less complex substances, so that your own system can utilize various aspects of the food where it may be needed in your body.

It is not unlike the processes involved in the construction of a building. The materials used for building have to be created out of trees, rocks and raw metals. The materials used in body building have to be created out of proteins and carbohydrates and fats.

When proteins enter the body, they do so in large chunks. These get broken down into the amino acids. In order to be used in the building of the body, the carbohydrates are broken down into sugars. And the fats are split up into fatty acids.

After this breaking down—digestive—process takes place, the materials necessary for building are ready. These are the foundation necessary for the next step of metabolism, the building up.

But, again, as in the construction of a building, you don't build at the place where you turned the trees into lumber, or shaped the rocks into building stones, or refined and molded the metal. So the substances that digestion has created are now absorbed into the system and are transported where they are most needed in the building process.

The amino acids and the sugars get into the blood capillaries. They are dispatched all over the body to feed the cells. The product of fat digestion, however, often proves to be the weak link in the building process. The tiny fat globules which have resulted from the digestive process of consumed fat re-combine and move about the body in what is known as the lymphatic system. While the blood carries the other products to their jobs quickly, the lymph—a fluid similar to blood plasma and which fills the spaces between the tissue cells—acts like stagnant water in a swamp. It has little or no movement. It is obvious, then, that the fat becomes congested around the body as these fatty deposits are laid down in layers under the skin or in more remote recesses.

Protein is, therefore, an essential building block for a

healthy body. And a fat body, obviously, is not a healthy body.

It is apparent, then, that there are certain essentials in maintaining a healthy diet—and a high-protein diet will help you lose weight healthfully and comfortably.

* * * * *

Rx *for Reducing:*

1. **Protein is the "first" food of life.**
2. **Certain basic foods can provide all the protein essentials.**
3. **Excessive intake of protein does not easily convert to fat.**
4. **Protein is essential to a healthy diet.**

* * * * *

5: A HIGH-PROTEIN DIET

If digestion were not automatic—in other words, if it took just one-tenth as much thinking to put food through the digestive system as it does to earn it—you would be far more careful of what you ate.

The chances are you wouldn't keep putting more cans of fat on shelves that are already overcrowded. You wouldn't neglect the effective and wonderfully complex vitamins that Nature provides in foods, and then attempt to make up deficiencies with the manufactured versions which modern science has perfected.

You would, obviously, eat sensibly. You would learn something about nutrition—the science of nourishing the body properly, providing it with the proteins, vitamins and minerals needed for its proper growth, maintenance and repair.

Yet, despite the fact that millions of words have been written to explain why you get fat and what you should do about it, there are still fifty-five million people in this country who weigh more than they should. Why? Simply because they eat too much food, or they eat too much of the wrong food.

It is as simple as that. Science has swept away all other theories and we are back to the original, basic reasoning. That is, if you feed your body more fuel than you burn up in energy, you store the excess in the form of fat.

Why do you eat too much? Well, there are many reasons—and this is probably not the place to go into them in detail. However, let us skim over some of the surface reasons. Most overeating, medicine has found, stems from a need to compensate for lack in some other area of life. Eating, then, can become a nervous habit. Or it can become a pleasure to substitute for other lack of pleasure. It can be an outlet for frustration. It can provide pleasant moments for someone who is lonely or unloved or discontented.

The list can go on and on, but the many reasons don't help with the solution of how to lose weight and keep that weight off. Essentially, when someone eats too much of the wrong food it is because he doesn't know better. There is no doubt that the help of a doctor in understanding one's own problem is a basic step to a complete cure. But most overweight people are not prepared to seek such help, or are unable to.

Nonetheless, overweight is a condition that can be prevented or corrected. And since overweight is caused by overeating, the only way to lose weight is to eat less.

As was pointed out earlier, many methods of dieting have been advocated. Any of these can prove to be successful if the dieting patient is motivated strongly enough or has enough insight into the problems that cause the overeating. But this is the most desirable and scientific way—the most ideal way—in which weight should be lost. Unfortunately, it is not practical for ninety-five per cent or more of those people who suffer from excessive weight.

The typical person who is overweight is not going to count calories, nor is he going to measure portions, nor, in fact, is he going to observe religiously for more than a few weeks at a time the complicated regime necessary for losing weight by calorie-counting. Most calorie-counting methods allow for "cheating." It's always possible, when you're totaling calories on a daily basis and adding them up on a weekly basis, to think that you're going to "make it up" tomorrow or the next day. In other words, today you can steal a forbidden sweet; tomorrow, you are positive, you will eat that many fewer calories. But it isn't that easy, and little by little the overindulgences, which are never made up, grow to the point where it's not just a forbidden sweet between meals but forbidden foods at mealtime themselves.

It's much like the problem of the alcoholic who always says or thinks that "one drink more" won't hurt. And the cured alcoholic has to avoid the "first drink" because that drink will get him back onto the alcohol trail once again.

The diet recommended, therefore, is one which has proved successful where calorie-counting diets have failed.

It is based on the principles of Dr. Leonid Kotkin and is expounded in his book, *Eat, Think and Be Slender,* first published in 1954.

The Kotkin Diet can be called an "exclusively high-protein, low-carbohydrate and no-fat" diet. It permits great quantities of foods high in nutritive content, and generally forbids foods which are low in these qualities. Practically all the foods which are rich in vitamins, amino acids and minerals are allowed. And foods which are rich in easily available calories are forbidden.

The theory is that people who are overweight have a great number of calories stored in the excess fat in their bodies and they cannot get rid of this extra weight unless these calories are used up.

The principle, then, is to utilize for daily energy needs the calories which are already stored in the body.

One of the best aspects of the diet is that there is no "starvation" necessary. The quantities of foods permitted on the diet are unlimited. In other words, *you can eat as much as you want of all the foods permitted.* There are never any quantity restrictions.

A problem that often is propounded by dieters is: "How will I get enough calories so that I will have sufficient energy?"

The answer to that is simple. The energy calories you need as you continue on your weight-reduction diet are supplied by your own body. Your body, filled with excess and unnecessary fatty tissue, contains countless calories which were taken from food indulged in the past and stored, in Nature's economical fashion, as fat.

* * * * *

Rx *for Reducing:*

1. **You are overweight because you overeat.**
2. **Excess food is converted and stored in the body as fat.**
3. **Most dieters "cheat" by eating more than is called for in their diets.**
4. **The high-protein diet helps the body burn up excessive stored fat without adding new fat—even if you cheat.**

* * * * *

Part Two: **DO IT!**

6: WHAT FOODS CAN YOU EAT?

As you have already learned in the first part of this book, the principle of this reducing diet is that of the "exclusively high-protein, low-carbohydrate and no-fat" theory. It permits great quantities of foods which are high in vitamins and nutritive content, and it generally forbids foods which are low in these qualities.

It has already been made quite clear that on this high-protein diet you can eat as much of the permitted foods as you want. On the other hand, foods which are rich in easily available calories are strictly forbidden. You have already learned that, being overweight, you have a great number of calories which are stored in your excess fat and you cannot get rid of your excess weight unless these calories are used up.

Therefore, you will be permitted to eat all lean meats, poultry and sea foods you desire. However, all the visible fat must be trimmed from these foods.

You will be permitted to eat all vegetables with five exceptions. The exceptions are the most common of the starchy vegetables. These are potatoes, corn, peas, rice, and beans of all types except string beans.

You will be permitted to use all condiments. There is no reason why salt (or, for that matter, water) should be limited in your diet, unless your physician finds that there is a medical reason for this.

You may eat snacks between meals. These should consist of variations of the permitted foods and nothing else.

This makes for a strict but simple diet and one which

can be adhered to easily. If you follow this diet, there is no possibility whatsoever of cheating.

But, to make it quite clear as to what you may not eat, let us outline the forbidden foods for your diet.

You may *not* eat bread or butter. A slice of protein bread is permissible at breakfast, but it should be omitted if possible.

You may *not* eat fruit, except half of a small grapefruit or a small orange at breakfast, and one-eighth of a small melon for dessert at dinner.

You may *not* eat sugars, fats or starches.

You may *not* eat breakfast cereals.

You may *not* have soup, except clear bouillon, consommé or other similar types.

You may *not* drink fruit juices, but vegetable juices are acceptable.

You may *not* drink liquor of any kind, or soft drinks. If you want to join your business lunch guests in a drink, order a "Bloody Awful"—which is a "Bloody Mary" without the vodka. It makes a sharp pick-me-up and aperitif, as does "bouillon on the rocks." You may drink tea and coffee in unlimited quantities—without sugar or cream (though skimmed milk and any of the artificial sweetening agents may be used).

You may *not* use oil or mayonnaise, but ketchup and cocktail sauce can be used in small quantities, or lemon juice or vinegar may be used as much as you want.

It is wise to remember that tinned and frozen foods are packed under scientific control by the most modern methods. They are wholesome, sanitary, thoroughly cooked and easily digestible. They retain their vitamin and mineral content to a high degree.

Moderation should be exercised in the intake of vegetables in a raw state. If eaten raw in excessive amounts, these vegetables may prove irritating. It's advisable to balance your vegetable intake by making sure that at least half of those you eat are cooked.

Don't think that this is a meager diet. It is far from it. Not only is there a wide variety of foods permissible, but you can eat as much of these permitted foods as you want. In a later chapter you will be given a great selection of menus and recipes that will show how easy it is to live on the high-protein diet, and eat well while losing weight.

It is interesting to note that the foods on this high-protein diet are, according to the National Research Council's Committee on Foods and Nutrition, the very foods that everyone needs in his everyday life. These are the foods most likely not to be stored as fat, even if we eat them in excessive quantities.

Even if you were to check these foods against a calorie chart, you would find that in most instances they contain fewer calories than the foods that are not permitted. But, remember, you are not going to count calories—calorie-counting becomes unnecessary when you follow a high-protein diet.

Let us examine from a nutritional point of view the basic foods on the diet.

Milk comes closer to being a perfect food, all by itself, than any other. It is a *must* in any diet—for babies, children, adolescents, and adults of all ages. It is one of the best sources of protein, and its goodly store of calcium makes strong bones and teeth. If you don't care too much for the taste of milk as a beverage, there are many ways of working it into a diet. You can add it to your coffee or tea. You can use it to mix with bouillon. However you use it, you should try to drink at least two cups (one pint) as a minimum each day. Skimmed milk has all the elements of highest benefit in milk with the exception that the butter-fat is removed.

While all cheeses are high in protein, many are also high in fat content. That is the reason that only *pot cheese* is recommended in the diet.

One of the best sources for protein intake—and also one of the best ways of filling yourself up—is by having good large portions of *meat* or *fish* or *poultry*. Lean beef or veal or lamb can supply you with many of the protein building blocks of the body, while keeping low the fat intake. Such variety meats as liver, heart and kidney are excellent sources of protein with none of the problems of fat intake.

Most fish—except for the naturally oily fishes such as bluefish—and all shellfish are extremely high in protein content. Chicken and turkey also fall into this category. Duck, on the other hand, is highly fatty and should be avoided.

You will notice that the diet calls for only one *egg* a day—and the egg should be boiled, not fried. This is because there is in eggs almost as much fat per gram as there is protein. Nonetheless, the mineral and vitamin value of eggs is so great that the extra fat obtained from one egg a day is not enough to upset the balance of the diet.

Among the *vegetables,* the leafy green and yellow ones supply minerals and vitamins, and they are an especially good source of Vitamin A, which is so valuable in order to keep the skin in good condition. Vitamin A increases resistance to colds and other infections, it protects eyesight and helps prevent night blindness. The body can store Vitamin A, so at least one serving a day—or more if you like—of the following vegetables should be taken:

Green asparagus, string or snap beans, broccoli, green celery, endive, escarole, lettuce and all salad greens, okra, peppers, spinach (and other greens such as chard, collard, kale, mustard and turnip), and carrots.

Among the white and root vegetables, which are a rich source of vitamins and minerals, you should also take at least one serving a day—or more if your hunger calls for it. These are to be boiled, or can be baked, but without fat, of course:

Artichoke, beets, cabbage, white celery, cauliflower (raw or cooked), cucumber, eggplant, mushrooms, onions, radishes, summer squash, and turnip.

Looking back, you can see that the high-protein diet leaves a great deal of leeway with regard to choice of menu, and even more leeway with regard to the amount you may eat

For the purposes of this book, it is assumed that the foods you eat have been harvested, processed, prepared, and served in such a manner that they meet the average values for the nutrient content listed in the table which follows on the succeeding pages.

This is a fairly safe assumption to make when calculating the average diet of large groups of people. We are, however, on more tenuous ground when we apply this to the individual person—or to the individual diet—simply because neither of these is an average.

However, if you follow the rules of diet outlined in this chapter, and elsewhere in this book, and if you are careful in the preparation of food to conserve the nutrients—as outlined in the next chapter—you can be assured that generally you will have a diet that is adequate from a nutritional point of view.

The table which follows is designed to show the composition of most of the common foods with regard to how much protein each contains as compared to the amount of fat and the amount of carbohydrate in each of these foods.

The list that is given here includes foods which are recommended on the high-protein diet as well as foods which have been specifically banned from the diet prescribed by this book. This is done especially to emphasize—as clearly as possible—why certain foods should be eaten, and—much more important—why the others should be avoided.

As you go down the list, you will notice which foods have a high protein content and what the content of protein is in relation to the amount of fat and carbohydrate contained in that particular food.

Where the content of fat and carbohydrate is low and the protein content is relatively higher, you know not only that you have a food which is permissible on the diet, but also that you have one which is advisable to add to your diet menu from the point of view of good nutrition.

Above all, remember that meals can be delicious, substantial, health-giving and, nevertheless, suitable for reducing—if the rules and the sample menus and recipes given in this book are followed faithfully.

It should be emphasized once again that water need not—nor should it—be limited while on a reducing diet. The adult normally needs about 2½ quarts of water every day, most of which is contained in the foods we eat. Remember, too, that your requirements for water increase with physical work or hot weather.

In the table that follows you will find a reference to the amount of fat included in the portions of meat listed. These references are noted with the small number 'T', the first of which you will find on page 49. There are such notations for other meat items, too.

The reference indicates that the outer layer of fat on the cut of meat in question was removed to within approximately one-half inch of the meat itself. That left enough fat to flavor the meat. However, for the purposes of this diet, all the fat should be removed from the outer side.

And while deposits of fat within the various cuts of

* * * * *

Rx ***for Reducing:***

1. **You may eat all you want of the permitted foods.**
2. **You must avoid completely all forbidden foods.**
3. **You may snack all you want with the permitted foods.**
4. **Study the following chart carefully now—and consult it frequently during your menu planning.**

* * * * *

Food, approximate measure		Pro-t-ein	Fat	Carbo-hy-drate
MILK, CREAM, CHEESE; RELATED PRODUCTS				
Milk, cow's:		*Grams*	*Grams*	*Grams*
Fluid, whole	1 cup	9	10	12
Fluid, nonfat (skim)	1 cup	9	Trace	13
Buttermilk, cultured, from skim milk.	1 cup	9	Trace	13
Evaporated, unsweetened, undiluted.	1 cup	18	20	24
Condensed, sweetened, undiluted.	1 cup	25	25	170
Dry, whole	1 cup	27	28	39
Dry, nonfat	1 cup	29	1	42
Milk, goat's: Fluid, whole	1 cup	8	10	11
Cream:				
Half-and-half (milk and cream).	1 cup	8	29	11
	1 tablespoon	Trace	2	1
Light, table or coffee	1 cup	7	52	10
	1 tablespoon	Trace	3	1
Whipping, unwhipped (volume about double when whipped):				
Medium	1 cup	6	78	8
	1 tablespoon	Trace	5	1
Heavy	1 cup	5	93	7
	1 tablespoon	Trace	6	Trace
Cheese:				
Blue mold (Roquefort type)	1 ounce	6	9	Trace
Cheddar or American:				
Ungrated	1-inch cube	4	6	Trace
Grated	1 cup	28	37	2
	1 tablespoon	2	2	Trace
Cheddar, process	1 ounce	7	9	Trace
Cheese foods, Cheddar	1 ounce	6	7	2

Food, approximate measure		Protein	Fat.	Carbohydrate
MILK, CREAM, CHEESE—Continued				
Cheese—Continued Cottage cheese, from skim milk:		*Grams*	*Grams*	*Grams*
Creamed	1 cup	30	11	6
	1 ounce	4	11	
Uncreamed	1 cup	38	1	6
	1 ounce	5	Trace	1
Cream cheese	1 ounce	2	11	1
	1 tablespoon	1	6	Trace
Swiss	1 ounce	7	8	1
Milk beverages:				
Cocoa	1 cup	9	11	26
Chocolate-flavored milk drink.	1 cup	8	6	27
Malted milk	1 cup	13	12	32
Milk desserts:				
Cornstarch pudding, plain (blanc mange),	1 cup	9	10	39
Custard, baked	1 cup	13	14	28
Ice cream, plain, factory packed:				
Slice or cut brick, 1/8 of quart brick.	1 slice or cut brick.	3	9	15
Container	3½fluid ounces.	2	8	13
Container	8 fluid ounces-	6	18	29
Ice milk	1 cup	9	10	42
Yoghurt, from partially skimmed milk.	1 cup	8	4	13
EGGS				
Eggs, large, 24 ounces per dozen:				
Raw:				
Whole, without shell	1 egg	6	6	Trace

White of egg	1 white	4	Trace	Trace
Yolk of egg	1 yolk	3	5	Trace
Cooked:				
Boiled, shell removed	2 eggs	13	12	1
Scrambled, with milk and fat.	1 egg	7	8	1
MEAT, POULTRY, FISH, SHELLFISH; RELATED PRODUCTS				
Bacon, broiled or fried crisp	2 slices	5	8	1
Beef, trimmed to retail basis,[1] cooked: Cuts braised, simmered, or pot-roasted:				
Lean and fat	3 ounces	23	16	0
Lean only	2.5 ounces	22	5	0
Hamburger, broiled:				
Market ground	3 ounces	21	17	0
Ground lean	3 ounces	23	10	0
Roast, oven-cooked, no liquid added:				
Relatively fat, such as rib:				
Lean and fat	3ounces	16	36	0
Lean only	1.8 ounces	14	7	0
Relatively lean, such as round:				
Lean and fat	3ounces	23	14	0
Lean only	2.5 ounces	21	4	0
Steak, broiled:				
Relatively fat, such as sirloin:				
Lean and fat	3ounces	20	27	0
Lean only	2 ounces	18	4	0
Relatively lean, such as round:				
Lean and fat	3 ounces	24	13	0
Lean only	2.4 ounces	22	4	0
Beef, canned:				
Corned beef	3 ounces	22	10	0
Corned beef hash	3 ounces	12	5	6
Beef, dried or chipped	2 ounces	19	4	0
Beef and vegetable stew	1 cup	15	10	15
Beef pot píe, baked: Individual pie,4¼-inch-diameter,weight before baking about 8 ounces.	1 pie	18	28	32
Chicken, cooked:				
Flesh and skin, broiled	3 ounces with out bone.	23	9	0

* Outer layer of fat on the cut was removed to within approximately ½ inch of the lean. Deposits of fat within the cut were not removed.

Food, approximate measure		Pro tein	Fat	Carbo-hy-drate
MEAT, POULTRY, FISH, SHELLFISH —Continued				
Chicken, cooked—Continued				
Breast, fried, ½ breast:		*Grams*	*Grams*	*Grams*
With bone	3.3 ounces	24	12	
Flesh and skin only	2.8 ounces	24	12	
Leg, fried (thigh and drumstick):				
With bone	4.3 ounces	27	15	
Flesh and skin only	3.1 ounces	27	15	
Chicken, canned, boneless	3 ounces	25	7	0
Chicken pot pie. Set Poultry potpie.				
Chile con carne, canned:				
With beans	1 cup	19	15	30
Without beans	1 cup	26	38	15
Heart, beef, trimmed of fat, braised.	3 ounces	26	5	1
Lamb, trimmed to retail basis,1 cooked:				
Chop, thick, with bone, broiled.	1 chop, 4.8 ounces.	25	33	0
Lean and fat	4 ounces	25	33	0
Lean only	2.6 ounces	21	6	0
Leg, roasted:				
Lean and fat	3 ounces	22	16	0
Lean only	2.5 ounces	20	5	0
Shoulder, roasted:				
Lean and fat	3 ounces	18	23	0
Lean only	2.3 ounces	17	6	0
Liver, beef, fried	2 ounces	13	4	6
Pork, cured, cooked:				
Ham, smoked, lean and fat	3 ounces	18	24	1
Luncheon meat:				
Cooked ham, sliced	2 ounces	13	13	0
Canned,spiced or unspiced	2 ounces,	8	14	1

Pork, fresh, trimmed to retail basis,[1] cooked:				
Chop, thick, with bone ___	1 chop, 3.5 ounces.	16	21	0
Lean and fat _________	2.3 ounces ___	16	21	0
L«an only .... _	1.7 ounces ___	15	7	0
Roast, oven-cooked, no liquid added:				
Lean and fat _________	3 ounces_____	21	24	0
Lean only ___________	2.4 ounces ___	20	10	0
Cuts simmered:				
Lean and fat _________	3 ounces_____	20	26	0
Lean only ___________	2.2 ounces __-_	18	6	0
Poultry potpië (chicken or turkey): Individual pie, 4¼-inch-diameter, about 8 ounces.	1 pie _______	17	28	39
Sausage:				
Bologna, slice 4.1 by 0.1 inch.	8 slices...........	27	62	2
Frankfurter,cooked _____	1 frankfurter..	6	14	1
Pork, bulk, canned ______	4 ounces_____	18	29	0
Tongue, beef, simmered ____	3 ounces ____	18	14	Trace
Turkey potpie♦ *See* Poultry potpie.				
Véal, cooked:				
Cutlet, broiled..--..... _	3 ounces with out bone.	23	9	0
Roast, medium fat, medium done: Lean and fat.	3 ounces.. -----	23	14	0
Fish and shellfish:				
Bluefish, baked or broiled-	3 ounces ____	22	4	0
Clams:				
Raw, meat only..............	3 ounces	11	1	3
Canned, solids and liquid.	3 ounces	7	1	2
Crabmeat, canned or cooked.	3 ounces ____	14	2	1
Fishsticks, breaded, cooked, frozen; stick, 3.8 by 1.0 by 0.5 inch.	10 sticks or 8-ounce package.	38	20	15
Haddock, fried..................	3 ounces	16	5	6
Mackerel:				
Broiled, Atlantic	3 ounces	19	13	0
Canned. Pacific. solids	3 ounces .	18	9	0

Food, approximate measure		Pro tein	Fat	Carbo-hy-drate
MEAT, POULTRY, FISH, SHELLFISH—Continued				
Fish and shellfish—Continued		*Grams*	*Grams*	*Grams*
Ocean perch, breaded (egg and breadcrumbs), fried.	3 ounces	16	11	6
Oysters, meat only: Raw, 13-19 medium selects.	1 cup	20	4	8
Oyster stew, 1 part oysters to 3 parts milk by volume, 3-4 oysters.	1 cup.	11	12	11
Salmon, pink, canned	3 ounces	17	5	0
Sardines, Atlantic type, canned in oil, drained solids.	3 ounces	22	9	1
Shad, baked	3 ounces	20	10	0
Shrimp, canned, meat only	3 ounces	23	1	—
Swordfish, broiled with butter or margarine.	3 ounces	24	5	0
Tuna, canned in oil, drained solids.	3 ounces	25	7	0
MATURE DRY BEANS AND PEAS, NUTS, PEANUTS; RELATED PRODUCTS				
Almonds, shelled	1 cup	26	77	28
Beans, dry:				
Common varieties, such 'as Great Northern, navy,' and others, canned:				
Red	1 cup	15	1	42
White, with tomato or molasses:				
With pork	1 cup	16	7	54
Without pork	1 cup	16	1	60
Lima, cooked	1 cup	16	1	48

Food	Measure			
Brazil nuts, broken pieces___	1 cup_______.	20	92	15
Cashew nuts, roasted______ .	1 cup____.... .	25	65	35
Coconut:				
Fresh, shredded.. ________	1 cup_.... ____	3	31	13
Dried, shredded, sweetened .	1 cup... _____	2	24	33
Cowpeas or blackeye peas, dry, cooked.	1 cup-.......	13	1	34
Peanuts, roasted, shelled:				
Halves ________________	1 cup... _____	39	71	28
Chopped ______ .... ___	1 tablespoon...	2	4	2
Peanut butter ____ .... ___	1 tablespoon_.	4	8	3
Peas, split, dry, cooked......	1 cup_.... _______	20	1	52
Pecans:				
Halves __________ . ____	1 cup_______	10	77	16
Chopped_______________ .	1 tablespoon..	1	5	1
Walnuts, shelled:				
Black or native, chopped...	1 cup......._	26	75	19
English or Persian:				
Halves______ .... ____	1 cup_-..........	15	64	16
Chopped____	1 tablespoon..	1	5	1
VEGETABLES AND VEGETABLE PRODUCTS				
Asparagus:				
Cooked, cut*8pears.......·	1 cup........	4	Trace	6
Canned spears, medium:				
Green _______________ .	6 spears _____	2	Trace	3
Bleached ....-	6 spears............	2	Trace	4
Beans:				
Lima, immature, cooked...	1 cup—......	8	1	29
Snap, green:				
Cooked:				
In small amount of water, short time.	1 cup... _______	2	Trace	6
In large amount of water, long time	1 cup........	2	Trace	6
Canned:				
Solids and liquid____ .	I cup _______	2	Trace	10
Strained or chopped __	1 ounce... ___	Trace	Trace	1
Bean sprouts, See Sprouts.				
Beets, cooked, diced................	1 cup__	2	Trace	16
Broccoli spears, cooked_____	1 cup_______	5	Trace	8
Brussels sprouts, cooked.....	1 cup........	6	I	12

Food, approximate measure		Pro tein	Fat	Carbo hy- drate
VEGETABLES—Continued				
Cabbage:				
Raw:		*Grams*	*Grams*	*Grams*
Finely shredded	1 cup	1	Trace	5
Coleslaw	1 cup	2	7	9
Cooked:				
In small amount of water, short time.	1 cup	2	Trace	9
In large amount of water, long time.	1 cup	2	Trace	9
Cabbage, celery or Chinese:				
Raw, leaves and stem, 1-inch pieces.	1 cup	1	Trace	2
Cooked	1 cup	2	1	5
Carrots:				
Raw:				
Whole, 5½ by 1 inch (25 thin strips).	1 carrot	1	Trace	5
Grated	1 cup	1	Trace	10
Cooked, diced	1 cup	1	1	9
Canned, strained or chopped.	1 ounce	Trace	0	2
Cauliflower, cooked, flower-buds.	1 cup	3	Trace	6
Celery, raw:				
Stalk, large outer, 8 by about 11/2 inches at root end.	1 stalk	1	Trace	1
Pieces, diced	1 cup	1	Trace	4
Collards, cooked	1 cup	7	1	14
Corn, sweet:				
Cooked, ear 5 by 1¾ inches	1 ear	2	1	16
Canned, solids and liquid	1 cup	5	1	41
Cowpeas, cooked, immature seeds.	1 cup	11	1	25

Cucumbers, 10-ounce; 7½ by about 2 inches:					
Raw, pared........................	1 cucumber.. .	1	Trace	6	
Raw, pared, center slice 1/8-inch thick.	*6* slices ____	Trace	Trace	1	
Dandelion greens, cooked____	1 cup ______	5	1	16	
Endive, curly (including escarole).	2 ounces ------	1	Trace	2	
Kale, cooked ____________	1 cup... ___	4	1	8	
Lettuce, headed, raw:					
Head, looscleaf, 4-inch-diameter.	1 head ______	3	3	Trace	6
Head, compact, 4¾-inch-diameter, 1 pound.	1 head ______	4	1	13	
Leaves	2 large or 4 small.	1	Trace	1	
Mushrooms,canned,solids and liquid.	1 cup _ . ___	3	Trace	9	
Mustard greens, cooked	1 cup ______	3	Trace	6	
Okra, cooked, pod 3 by 5/8 inch. Onions:	8 pods_____	2	Trace	6	
Mature:					
Raw, onîon 2½-ìnch-diameter.	1 onion... ___				
		2	Trace	11	
Cooked _____________	1 cup______				
Young green, small, without tops.	6 onions _____	2	Trace	18	
		Trace	Trace	5	
Parsley, raw, chopped _____	1 tablespoon_.				
Parsnips,cooked ---------------	1 cup..	Trace	Trace	Trace	
Peas, green:		2	1	22	
Cooked _______________	1 cup_______				
Canned, solids and liquid—	1 cup_-..........	8	1	19	
Canned, strained	1 ounce	8	1	32	
Peppers, hot, red, without	1 tablespoon...	1	Trace	2	
seeds, dried; ground chili powder. Peppers, sweet:		2	1	9	
Raw, medium, about 6 per poundîng					
Green pod without stem and seeds.	1 pod.....___	1	Trace	3	
Red pod without stem and seeds.	1 pod...____	*1*	Trace	4	
Canned,pimientos,medíum_	1 pod _______				
Potatoes, medium, about 3 per pound: Baked, peeled after baking.	1 potato......	Trace	Trace	2	
		3	Trace	21	

Food, approximate measure		Protein	Fat	Carbohydrate
VEGETABLES—Continued				
Potatoes—Continued				
Boiled:		*Grams*	*Grams*	*Grams*
Peeled after boiling	1 potato	3	Trace	23
Peeled before boiling	1 potato	3	Trace	21
French-fried, piece 2 by 1/2 by 1/2 inch:				
Cooked in deep fat, ready to eat.	10 pieces	2	7	20
Frozen, ready to heat for serving.	10 pieces	2	4	15
Mashed:				
Milk added	1 cup	4	1	30
Milk and butter added	1 cup	4	12	28
Potato chips, medium, 2-inch-diameter.	10 chips	1	7	10
Pumpkin, canned.	1 cup	2	11	8
Radishes, raw, small, without tops.	4radishes	Trace	Trace	2
Sauerkraut, canned, drained solids.	1 cup	2	Trace	7
Spinach:				
Cooked	1 cup	6	1	6
Canned, drained solids	1 cup	6	1	6
Canned, strained and creamed.	1 ounce	1	Trace	2
Sprouts, raw:				
Mung bean	1 cup	3	Trace	4
Soybean	1 cup	7	1	6
Squash:				
Cooked:				
Summer, diced	1 cup	1	Trace	8
Winter, baked, mashed.	1 cup	4	12	3
Canned, winter, strained or chopped.	1 ounce	Trace	Trace	2

Food	Measure			
Sweetpotatoes:				
Cooked, medium, 5 by 2 inches, weight raw 6 ounces:				
Baked, peeled after baking.	1 sweetpotato.	2	1	*36*
Boiled, peeled after boiling.	1 sweetpotato.	*2*	1	39
Candied, *3½ by 2¼* inches..	1 sweetpotato.	2	6	60
Canned, vacuum or solid pack.	1 cup ____.........	4	Trace	54
Tomatoes:				
Raw, medium, 2 by *2½:* , inches, about 3 per pound.	1 tomato______	2	Trace	6
Canned or cooked........... __	1 cup_______	2	Trace	9
Tomato juice, canned................	1 cup	2	Trace	10
Tomato catsup _______	1 tablespoon..	Trace	Trace	4
Turnips, cooked, diced _______	1 cup ______	1	Trace	9
Turnip greens:				
Cooked:				
In small amount of water, short time.	1 cup__,, ____	4	1	8
In large amount of water, long time.	1 cup _______	4	1	8
Canned,solids and liquid___	1 cup...............	3	1	7
FRUITS AND FRUIT PRODUCTS				
Apples,raw,medium,2½-inch-diameter, about 3 per pound.	1 apple________	Trace	Trace	18
Apple brown betty	1 cup	4	8	69
Applejuice, fresh or canned ___	1 cup_______	Trace	0	34
Applesauce, canned:				
Sweetened...................... ___	1 cup ______	Trace	Trace	50
Unsweetened	1 cup	Trace	Trace	26
Apricots:				
Raw, about 12 per pound..	3 apricots ___ .	1	Trace	14
Canned in heavy syrup:				
Halves and syrup _____._	1 cup................	2	Trace	57
Halves, medium, and syrup.	4 halves; 2 tablespoons syrup.	1	Trace	27
Dried:				
Uncooked, 40 halves, small.	1 cup________	8	1	100
Cooked, unsweetened, fruit and liquid.	**1** cup ______	5	1	62

Food, approximate measure		Prot eîn	Fat	Carbo-hy-drate
FRUITS—Continued				
		Grams	*Grams*	*Grams*
Apricots and applesauce, canned (strained or chopped).	1 ounce......	Trace	Trace	5
Apricot nectar...	1 cup........	1	Trace	36
Avocados, raw:				
California varieties, mainly Fuerte: 10-ounce avocado, about 3¼ by 4¼ inches, peeled, pitted	½ avocado....	2	18	6
½-inch cubes	1 cup	3	26	9
Florida varieties:				
13-ounce avocado, about 4 by 3 inches, peeled, pitted.	½avocado	2	14	11
½- inch cubes	1 cup	2	17	13
Bananas, raw, 6 by ½ inches, about 3 per pound.	1 banana.....	1	Trace	23
Blackberries, raw	1 cup	2	11	9
Blueberries,raw	1 cup	1	12	1
Cantaloups, raw, medium, 5-inch-diameter, about 1¾ pounds.	½ melon.....	1	Trace	9
Cherries:				
Raw, sour, sweet, hybrid—	1 cup	1	1	15
Canned, red, sour, pitted—	1 cup	2	12	6
Cranberry juice cocktail, canned.	1 cup	Trace	Trace	36
Cranberry sauce, sweetened, canned or cooked.	1 cup	Trace	1	142
Dates, "fresh" and dried, pitted, cut	1 cup	4	1	134

Food	Measure			
Figs:				
Raw, small, 1½-inch-diameter, about 12 per pound.	3 figs	2	Trace	22
Dried, large, 2 by 1 inch	1 fig	1	Trace	15
Fruit cocktail, canned in heavy syrup, solids and liquid.	1 cup	1	1	50
Grapefruit:				
Raw, medium, 4¼-inch-diameter, size 64:				
White	½ grapefruit	1	Trace	14
Pink or red	½ grapefruit	1	Trace	14
Raw sections, white	1 cup	1	Trace	20
Canned:				
Sirup pack, solids and liquid.	1 cup	1	Trace	44
Water pack, solids and liquid.	1 cup	1	Trace	18
Grapefruit juice:				
Fresh	1 cup	1	Trace	23
Canned:				
Unsweetened	1 cup	1	Trace	24
Sweetened	1 cup	1	Trace	32
Frozen, concentrate, unsweetened:				
Undiluted, can, 6 fluid ounces.	1 can	4	1	72
Water added	1 cup	1	Trace	24
Frozen, concentrate, sweetened:				
Undiluted, can, 6 fluid ounces.	1 can	3	1	85
Water added	1 cup	1	Trace	28
Dehydrated:				
Crystals, can, net weight 4 ounces.	1 can	5	1	103
Water added	1 cup	1	Trace	24
Grapes, raw:				
American type (slip skin), such as Concord, Delaware, Niagara, and Scuppernong.	1 cup	1	1	16
European type (adherent skin), such as Malaga, Muscat, Suìtanina (Thompson Seedless), and Flame Tokay.	1 cup	1	Trace	26
Grapejuice, bottled	1 cup	1	Trace	42

Food, approximate measure		Pro-tein	Fat	Carbo-hy-dratc
FRUITS—Cóntinued		*Grams*	*Grams*	*Grams*
Lemons, raw, medium, 21/8-inch-diameter, size 150.	1 lemon......	1	Trace	6
Lemon juice:				
Fresh....................	1 cup........	1	Trace	20
	1 tablespoon..	Trace	Trace	1
Canned,unsweetened. __..	1 cup_	1	Trace	19
Lemonade concentrate, frozen, sweetened:				
Undiluted, can, 6 fluid ounces.	1 can........	Trace	Trace	112
Water added _____	1 cup........	Trace	Trace	28
Lime juice:				
Fresh_________________	1 cup_______	1	Trace	22
Canned........ ________	1 cup.........	1	Trace	22
Limeade concentrate, frozen, sweetened:				
Undiluted, can, 6 fluid ounces.	1 can........	Trace	Trace	108
Water added__ . _______	1 cup... ____	Trace	Trace	27
Oranges, raw: •				
Navel, California (winter), size 88, 2½inch-diameter.	1 orange___ ...	2	Trace	16
Other varieties, 3-inch-diameter.	1 orange... __	1	Trace	18
Orange juîce:				
Fresh:				
California, Valencia, summer· 'Florida	1 cup........	2	1	26
varieties:				
Early and midseason _	1 cup........	1	Trace	23
Late season, Valencia..	1 cup_______	1	Trace	26
Canned, unsweetened -------	1 cup....-----	2	Trace	28
Frozen concentrate:				
Undiluted, can, 6 fluid ounces.	1 can......	5	Trace	80

Water added	1 cup	2	Trace	27
Dehydrated:				
Crystals, can, net weight 4 ounces.	1 can	6	2	100
Water added	1 cup	1	Trace	27
Orange and grapefruit juice:				
Frozen concentrate:				
Undiluted, can, 6 fluid ounces.	1 can	4	1	78
Water added	1 cup	1	Trace	26
Papayas, raw, ½-inch cubes	1 cup	1	Trace	18
Peaches:				
Raw:				
Whole, medium, 2-inch-diameter, about 4 per pound.	1 peach	1	Trace	10
Sliced	1 cup	1	Trace	16
Canned, yellow-fleshed/solids and liquid:				
Syrup pack, heavy:				
Halves or slices	1 cup	1	Trace	52
Halves, medium, and syrup.	2 halves and 2 tablespoons syrup.	Trace	Trace	24
Water pack	1 cup	1	Trace	20
Strained	1 ounce	Trace	Trace	5
Dried:				
Uncooked	1 cup	5	1	109
Cooked, unsweetened, 10-12 halves and 6 tablespoons liquid.	1 cup	3	1	58
Frozen:				
Carton, 12 ounces	1 carton	1	Trace	69
Can, 16ounces	1 can	2	Trace	92
Peach nectar, canned	1 cup	Trace	Trace	31
Pears:				
Raw, 3 by 2½-inch-diameter.	1 pear	1	1	25
Canned, solids and liquid:				
Syrup pack, heavy:				
Halves or slices	1 cup	1	1	50
Halves, medium, and syrup.	2 halves and 2 tablespoons syrup.	Trace	Trace	23

Food, approximate measure		Protein	Fat	Carbohydrate
FRUITS—Continued				
Pears, canned, solids and liquid—Continued		*Grams*	*Grams*	*Grams*
Water pack	1 cup	Trace	Trace	20
Strained	1 ounce.	Trace	Trace	4
Pear nectar, canned	1 cup	1	Trace	33
Persimmons, Japanese or Kaki, raw, seedless, 2½-inch-diameter.	1 persimmon..	1	Trace	20
Pineapple:				
Raw, diced	1 cup	1	Trace	19
Canned, syrup pack, solids and liquid:				
Crushed	1 cup	1	Trace	55
Sliced, slices, and juice	2 small or 1 large and 2 tablespoons juice.	Trace	Trace	26
Pineapple juice, canned	1 cup	1	Trace	32
Plums, all except prunes:				
Raw, 2-inch-diameter, about 2 ounces.	1 plum	Trace	Trace	7
Canned, syrup pack (Italian prunes):				
Plums and juice	1 cup	1	Trace	50
Plums (without pits) and juice.	3 plums and 2 tablespoons juice.	Trace	Trace	25
Prunes, dried, Medium, 50-60 per pound:				
Uncooked	4 prunes	1	Trace	19
Cooked, unsweetened, 17-18 prunes and 1/3 cup liquid.	1 cup	3	1	81
Canned, strained	1 ounce	Trace	Trace	7
Prune juice, canned	1 cup	1	Trace	45
Raisins, dried	1 cup	4	Trace	124

Raspberries, red:				
Raw.........	1 cup............	1	1	17
Frozen, 10-ounce carton __	1 carton____	2	1	70
Rhubarb, cooked, sugar added.	1 cup ______	1	Trace	98
Strawberries:				
Raw, capped.. _________	1 cup______	1	1	13
Frozen, 10-ounce carton __	1 carton____	2	1	75
Frozen, 16-ounce can ____	1 can ______	1	2	121
Tangerines, raw, medium, 2½-inch-diameter, about 4 per pound.	1 tangerine__	1	Trace	10
Tangerine juice:				
Canned, unsweetened ____	1 cup_______	1	Trace	25
Frozen concentrate:				
Undiluted, can, 6 fluid ounces.	1 can.............	4	1	8O
Water added _________	1 cup_______	1	Trace	27
Watermelon, raw, wedge, 4 by 8 inches (Me of 10- by 16-inch melon, about 2 pounds with rind).	1 wedge ____	2	1	29

GRAIN PRODUCTS

Barley, pearled, light, uncooked.	1 cup	17	2	160
Biscuits, baking powder, with enriched flour, 2½-inch-diameter.	1 biscuit..........	3	4	18
Bran flakes (40 percent bran) with added thiamine.	1 ounce _____	3	1	22
Breads:				
Boston brown bread, made with degermed cornmeal, slice, 3 by ¼ inch.	1 slice ______	3	1	22
Cracked-wheat bread:				
Loaf, 1-pound, 20 slices__	1 loaf ______	39	10	236
Slice____ .... ______	1 slice...........	2	1	12
French or vienna bread:				
Enriched, 1-pound loaf _	1 loaf__	41	14	251
Unenriched, 1-pound loaf.	1 loaf_______	41	14	251
Italian bread:				
Enriched, 1-pound loaf _	1 loaf______	41	4	256
Unenriched, 1-pound loaf_	1 loaf	41	4	256

Food, approximate measure		Protein	Fat	Carbohydrate
GRAIN PRODUCTS—Continued				
Breads—Continued				
Raisin bread:		*Grams*	*Grams*	*Grams*
Loaf, 1-pound, 20 slices	1 loaf	30	13	243
Slice	1 slice	2	1	12
Rye bread: American, light (*1/3* rye, 2/3 wheat):				
Loaf, 1-pound, 20 slices.	1 loaf	41	5	236
Slice	1 slice	2	Trace	12
Pumpernickel, dark, loaf, 1 pound.	1 loaf	41	5	241
White bread, enriched: [2]				
1 to 2 percent nonfat dry milk:				
Loaf,1-pound,20slices	1 loaf	39	15	229
Slice	1 slice	2	1	12
3 to 4 percent nonfat dry milk:				
Loaf, 1-pound	1 loaf	39	15	229
Slice, 20 per loaf	1 slice	2	1	12
Slice, toasted	1 slice	2	1	12
Slice, 26 per loaf	1 slice	1	1	9
5 to 6 percent nonfat dry milk:				
Loaf, 1-pound, 20 slices.	1 loaf	41	17	228
Slice	1 slice	2	1	12
White bread, unenriched:[2]				
1 to 2 percent nonfat dry milk:				
Loaf, 1-pound, 20 slices.	1 loaf	39	15	229
Slice	1 slice	2	1	12

[8] When the amount of nonfat dry milk in commercial white bread îs unknown, use values for bread with ¾ to 4 percent nonfat dry milk.

3 to 4 percent nonfat dry milk:				
Loaf,l-pound	1 loaf............	39	15	229
Slice,20perloaf ___	1 slice...........	2	1	12
Slice,toasted ___	1 slice______	2	1	12
Slice, 26 per loaf__	1 slice______	1	1	9
5 to 6 percent nonfat dry milk:				
Loaf, 1-pound, 20 slices.	î loaf..........	41	17	228
Slice _____________	1 slice_____	2	1	12
Whole-wheat, graham, entire-wheat bread:				
Loaf, 1-pound, 20 slices__	1 loaf ____	48	14	216
Slice _______________	1 slice_____	2	1	11
Toast........_ ....____	1 slice...........	2	1	11
Breadcrumbs, dry, grated __	1 cup............	11	4	65
Cakes:				
Angel food cake; sector, 2-inch (1/12 of 8-inch-diam-eter cake).	1 sector_____	3	Trace	23
Chocolate cake, fudge icing; sector, 2-inch (¼c of 10-inch-diameter layer cake).	1 sector___	5	14	70
Fruitcake, dark; piece, 2 by 2 by ¼ inch.	1 piece _____	2	4	17
Gingerbread; piece, 2 by 2 by 2 inches. Plain cake and cupcakes, without icing:	1 piece	2	7	28
Piece, 3 by 2 by 1½ inches.	1 piece ___	4	5	31
Cupcake, 2¾-inch-diam-eter. Plain cake and cupcakes, with icing:	1 cupcake _	3	3	23
Sector, 2-inch 1/16 of 10-inch layer cake).	1 sector____	5	6	62
Cupcake, 2¾-inch-diame-ter.	1 cupcake __	3	3	31
Pound cake; slice, 2¾ by 3 by 5/8 inch.	1 slice	2	7	15
Sponge cake; sector, 2-inch (1/13 of 8-inch¯diameter cake). Cookies:	1 sector ____	3	2	22
Plain and assorted, 3-inch-diameter.	1 cooky _____	2	3	19

Food, approximate measure		Protein	Fat	Carbohydrate
GRAIN PRODUCTS—Continued				
Cookies—Continued		*Grams*	*Grams*	*Grams*
Fig bars, small	1 fig bar	1	1	12
Corn-cereal mixture (mainly degermed cornmeal), puffed, with added thiamine, niacin, and iron.	1 ounce	2	1	23
Corn flakes, with added thiamine, niacin, and iron:				
Plain	1 ounce	2	Trace	24
Presweetened	1 ounce	1	Trace	26
Corn grits, white, degermed, cooked:				
Enriched	1 cup	3	Trace	27
Unenriched	1 cup	3	Trace	27
Cornmeal, white or yellow, dry:				
Whole ground	1 cup	11	5	87
Degermed, enriched	1 cup	11	2	114
Corn muffins, made with enriched, degermed cornmeal; muffin, 2¾-inch diameter.	1 muffin	4	5	22
Corn, puffed, presweetened, with added thiamine, riboflavin, niacin, and iron.	1 ounce	1	Trace	26
Corn and soy shreds, with added thiamine and niacin.	1 ounce	5	Trace	21
Crackers:				
Graham	4 small or 2 medium.	1	1	10
Saltines, 2 inches square	2 crackers	1	1	6
Soda, plain:				
Cracker, 2½ inches square	2 crackers	1	1	8
Oyster crackers	10 crackers	1	1	7
Cracker meal	1 tablespoon	1	1	7
Doughnuts, cake type	1 doughnut	2	7	17

Farîna, cooked; enriched to minimum levels for required nutrients and for the optional nutrient, calcium.	1 cup	3	Trace	22
Macaroni, cooked:				
Enriched:				
Cooked 8-10 minutes (undergoes additional cooking in a food mix ture).	1 cup	6	1	39
Cooked until tender	1 cup	5	1	32
Unenriched:				
Cooked 8-10 minutes (un-dergoes additional cook-ing in a food mixture).	1 cup	6	1	39
Cooked until tender	1 cup	5	' 1	32
Macaroni, enriched, and cheese, baked.	1 cup	18	25	44
Muffins, with enriched white flour; muffin, 2¾-inch-diam-cter.	1 muffin	4	5	19
Noodles (egg noodles), cooked:				
Enriched	1 cup	7	2	37
Unenriched	1 cup	7	2	37
Oat-cereal mixture,mainly oats, with added B-vitamins and minerals.	1 ounce-,	4	2	21
Oatmeal or rolled oats, regular or quick-cooking, cooked.	1 cup.-	5	3	26
Pancakes (griddlecakes),4-inch-diameter:				
Wheat, enriched flour (home recipe).	1 cake	2	2	8
Buckwheat (buckwheat pancake mix).	1 cake	2	2	6
Piecrust, plain, baked: Enriched flour:				
Lower crust, 9-inch shell _	1 crust	10	36	72
Double crust, 9-inch pie_	1 double crust_	20	73	143
Unenriched flour:				
Lower crust, 9-inch shell.	1 crust	10	36	72
Double crust, 9-inch pie_.	1 double crust.	20	73	143

Food, approximate measure		Protein	Fat	Carbohydrate
GRAIN PRODUCTS—Continued				
Pies; sector, 4-inch, 1/2 of 9-inch-diameter pie:		*Grams*	*Grams*	*Grams*
Apple	1 sector	3	13	S3
Cherry	1 sector	3	13	55
Custard	1 sector	7	11	34
Lemon meringue	1 sector	4	12	45
Mince	1 sector	3	9	62
Pumpkin	1 sector	5	12	34
Pizza (cheese), 5 ½-inch sector, *1/8* of 14-inch-diameter pie.	1 sector	8	6	23
Popcorn, popped	1 cup	2	1	11
Pretzels, small stick	5 sticks	Trace	Trace	4
Rice, cooked:				
Parboiled	1 cup	4	Trace	45
White	1 cup	4	Trace	44
Rice, puffed, with added thiamine, niacin, and iron.	1 cup	1	Trace	12
Rice flakes, with added thiamine and niacin.	1 cup	2	Trace	26
Rolls:				
Plain, pan; 12 per 16 ounces:				
Enriched	1 roll	3	2	20
Unenriched	1 roll	3	2	20
Hard, round; 12 per 22 ounces.	1 roll	5	2	31
Sweet, pan; 12 per 18 ounces.	1 roll	4	4	21
Rye wafers, 1⅞ by 3½ inches	2 wafers	2	Trace	10
Spaghetti, cooked until tender:				
Enriched	1 cup	5	1	32
Unenriched	1 cup	5	1	32

Spaghetti with meat sauce___	1 cup_______	13	10	35
Spaghetti in tomato sauce with cheese.	1 cup_______	6	5	36
Waffles, with enriched flour, ½ by 4½by5½inches.	1 waffle_____	8	9	30
Wheat, puffed:				
With added thiamine, niacin, and iron.	1 ounce	4	Trace	22
With added thiamine and niacin; presweetened.	1 ounce _____	1	Trace	26
Wheat, rolled; cooked	1 cup..............	5	1	40
Wheat, shredded, plain (long, round, or bite-size).	1 ounce--------	3	1	23
Wheat and malted barley cereal, with added thiamine, niacin, and iron.	1 ounce _____	3	Trace	24
Wheat flakes, with added thiamine, niacin, and iron.	1 ounce	3	Trace	23
Wheat flours:				
Whole-wheat, from hard wheats, stirred.	1 cup_______	16	2	85
All-purpose or family flour:				
Enriched, sifted..............	1 cup ---------	12	1	84
Unenriched, sifted........ -	1 cup............_	12	1	84
Self-rising:				
Enriched............______	1 cup ---------	10	1	81
Unenriched....................	1 cup._	10	1	81
Wheat germ, stirred	1 cup.............	17	7	34

FATS, OILS

Butter, 4 sticks per pound:				
Sticks, 2	1 cup_	1	181	1
Stick, ¼	1 tablespoon..	Trace	11	Trace
Patorsquare(64perpound)-	1 pat.............	Trace	6	Trace
Fats, cooking:				
Lard_________________	1 cup ______	0	220	0
	1 tablespoon__	0	14	0
Vegetable fats__________	1 cup__.........	0	200	0
	1 tablespoon. _	0	12	0

Food, approximate measure		Prot ein	Fat	Carbo-hy-drate
FATS, OILS-Continued				
Margarine, 4 sticks per pound:		*Grams*	*Grams*	*Grams*
Sticks, 2..	1 cup	1	181	1
Stick, ½---	1 tablespoon..	Trace	11	Trace
Pat or square (64 per pound).	1 pat	Trace	6	Trace
Oils, salad or cooking:				
Corn	1 tablespoon..	0	14	0
Cottonseed	1 tablespoon..	0	14	0
Olive	1 tablespoon..	0	14	0
Soybean	1 tablespoon..	0	14	0
Salad dressings:				
Blue cheese	1 tablespoon..	1	101	
Commercial, plain; mayonnaise type.	1 tablespoon.	Trace	62	
French	1 tablespoon..	Trace	6	2
Home cooked, boiled	1 tablespoon.	1	2	3
Mayonnaise	1 tablespoon..	Trace	12	Trace
Thousand Island	1 tablespoon	Trace	8	1
SUGARS, SWEETS .				
Candy:				
Caramels	1 ounce	1	3	22
Chocolate, sweetened, milk.	1 ounce	2	91	6
Fudge, plain	1 ounce	Trace	32	3
Hard candy	1 ounce	0	0	28
Marshmallow	1 ounce	1	0	23
Chocolate syrup	1 tablespoon.	Trace	Trace	11
Honey, strained or extracted	1 tablespoon	Trace	0	17
Jams,marmalades,preserves	1 tablespoon..	Trace	Trace	14
Jellies	1 tablespoon.-	0	0	13

Molasses, cane:				
Light (first extraction) ____	1 tablespoon __	....	....	13
Blackstrap(third extraction)	1 tablespoon__	----- .	....	11
Syrup, table blends ________	1 tablespoon. _	0	0	15
Sugar:				
Granulated, cane or beet __	1 cup ______	0	0	199
	1 tablespoon__	0	0	12
	1 lump _____	0	0	7
Powdered, stirred before measuring.	1 cup_______	0	0	127
	1 tablespoon__	0	0	8
Brown, firm-packed______	1 cup ______	0	0	210
	1 tablespoon. _	0	0	13
MISCELLANEOUS ITEMS				
Beer (average 4 percent alcohol). Beverages,	1 cup ______	1	Trace	11
carbonated:				
Ginger ale _____________	1 cup ______	______	______	21
Kola type _____________	1 cup ______	.— ...	....	28
Bouillon cube, 5/8 inch _____	1 cube _____	Trace	Trace	0
Chili powder. *See* Vegetables, Peppers.				
Chili sauce (mainly tomatoes) _	1 tablespoon. _	Trace	Trace	4
Chocolate:				
Bitter or unsweetened ____	1 ounce ____	2	15	8
Sweetened _____________	1 ounce ____	1	8	18
Cider. *See* Fruit, Applejuice.				
Gelatin, dry:				
Plain _________________	1 tablespoon..	9	Trace	0
Dessert powder, 3-ounce package. Gelatin dessert,	½ cup______	8	Trace	76
ready-to-eat:				
Plain _________________	1 cup ______	4	Trace	36
With fruit..........................	1 cup	3	Trace	42
Olives, pickled:				
Green__ .... _____	12 Extra Large or 7Jumbo.	1	7	1
Ripe: Mission;other varieties, such as Ascolano, Manzanillo, and Sevillano.	12Extra Large or 7 Jumbo.	1	9	2

Food, approximate measure		Protein	Fat	Carbohydrate
MISCELLANEOUS ITEMS—Continued				
Pickles, cucumber:		*Grams*	*Grams*	*Grams*
Dill, large, 4 by 1¾ inches..	1 pickle..............	1	Trace	3
Sweet, *23/4* by 3/4 inch..........	1 pickle............	Trace	Trace	5
Popcorn. *See* Grain Products.				
Sherbet, factory packed ____	1 cup_______	3	Trace	58
Soups, canned; ready-to-serve:				
Bean ....,..............	1 cup	8	5	30
Beef.......................................	1 cup................	6	4	11
Bouillon, broth, consommé	1 cup.................	2	....	0
Chicken________________	1 cup_______	4	2	10
Clam chowder __________	1 cup ...._	5	2	12
Cream soup (asparagus, eel-ery, mushroom)	1 cup	7	12	18
Noodle, rice, barley ______	1 cup_______	6	4	13
Pea..	1 cup................	6	2	25
Tomato____ .... ____	1 cup_______	2	2	18
Vegetable ______ __	1 cup_____________	4	2	14
Starch, pure, including/arrow-root, corn, etc.	1 cup................	1	Trace	111
	1 tablespoon. _	Trace	Trace	7
Tapioca, quick-cooking granulated, dry; stirred before measuring	leup................	1	Trace	131
—	1 tablespoon_.	Trace	Trace	8
Vinegar..........	1 tablespoon..	0	....	1
White 6auce, medium_____...	1 cup...............	10	33	23
Yeast:				
Baker's:				
Compressed... __ ____	1 ounce _____	3	Trace	3
Dry active ___________	1 ounce _____	10	Trace	11
Brewer's, dry	1 tablespoon..	3	Trace	3
Yoghurt. *See* Milk, Cream, Cheese; Related Products.				

7: COOKING TO SAVE NUTRITION

Although it may never have occurred to you, nutrition is a highly personal matter. In fact, it is quite as personal as your diary or, perhaps, your income-tax report.

Your nutrition actually can determine a great deal about your everyday life—how you look, how you act, how you feel. Whether you are grouchy or cheerful can depend upon your nutrition. Whether you are homely or beautiful can, to a great extent, depend upon your nutrition. Whether you think clearly or in a confused manner, whether you enjoy your work or make a drudgery of it, whether you increase your power to earn or stay in an economic rut—all these things can, surprisingly enough, depend upon your nutrition.

So you see, the foods that you eat daily can make the difference between your living "the good life," or constantly fighting an uphill battle, much as the legendary King Sisyphus, who was condemned forever to rolling a stone toward the top of a hill only to have it roll down again before he reached the summit.

What is nutrition? This is a question many people fail to ask simply because they know the word and are inclined to feel that knowing the word means understanding the concept. Simply, nutrition is the study of how foods, after you take them, affect you. It should not be confused with dietetics. Dietetics is the study of foods which you should eat.

Unfortunately, food faddists and crackpots have given nutrition some bad times. These people—and we will be discussing food fads later on in the book—usually have no scientific training. They are responsible for disseminating huge amounts of misinformation. They usually make claims which can't be justified and, as so often occurs in the case of the faddist and crackpot, they are only out for commercial gain, not to help you. The only thing they succeed in doing, aside from making money for themselves, is to make any thinking person skeptical about the whole subject of nutrition.

In the field of diet nutrition there are a flock of faddists. Some remain in the forefront for years, others achieve short and spectacular success and then fade away much as a comet shooting through the sky.

One of the unfortunate facts about the nutritionist faddists is that they mislead people into diets that often are well-intentioned but are badly followed. For example, there are a number of food crackpots who talk of a high-protein diet, just as does this book. But there is no attempt to balance the diet and it has been discovered by medical men investigating the diet of patients who claimed to have been following such a regime that their protein intake was often as little as one-third of that recommended by the National Research Council.

To give you a basic idea of what your daily intake of protein in grams should be, here is the National Research Council's recommended daily allowance, based on the same categories as the caloric intake chart earlier in this book:

	AGE	WEIGHT	HEIGHT	PROTEIN IN GRAMS
MEN	25	154	69 inches	70
	45	154	69 inches	70
	65	154	69 inches	70
WOMEN	25	128	64 inches	58
	45	128	64 inches	58
	65	128	64 inches	58
Pregnant (last 4½ months)				+20
Breast feeding (850 ml. daily)				+40

	AGE	WEIGHT	HEIGHT	PROTEIN IN GRAMS
BOYS	13-15	108	64 inches	85
	16-19	139	69 inches	100
GIRLS	13-15	108	63 inches	80
	16-19	120	64 inches	75
CHILDREN	1-3	27	34 inches	40
	4-6	40	43 inches	50
	7-9	60	51 inches	60
	10-12	79	57 inches	70

The availability of the ingredients of a good diet does not necessarily assure you of a good diet. What matters is the food that you actually eat. There are great losses between the farm where the food is grown, the store where it is sold, and the table at which you eat it. Your selection of foods and the way you store and cook them are important.

Since few people today do their own canning or freezing we will not go into the nutritional losses that foods suffer in preserving and storing. Instead, we will limit this particular discussion to the nutritional losses that can happen in cooking, and how the vitamin and mineral values of the foods you want on a high-protein diet can be preserved to the greatest amount possible while they are being cooked.

Of those nutrients which are best known, the ones most subject to destruction during cooking are: folic acid, Vitamin A, thiamine, and riboflavin. Let us, before we go ahead, look at these nutrients and see where they come from and what they do.

Folic acid is a member of the B-group of vitamins. It is present in small amounts in most common foods, but the richest sources of folic acid are deep green leafy vegetables and meats, particularly liver. Folic acid is essential for the metabolism of growing tissues and cells.

And because the losses of folic acid in the cooking of foods can be considerable, it is advisable that vegetables be stored under refrigeration to maintain as high as possible a folic acid content before cooking.

Ascorbic acid is known as Vitamin C. Most fresh plant

foods contain some ascorbic acid and citrus fruits are, particularly, high in it. It is also found in tomatoes, peppers (green or red), cabbage, cauliflower, broccoli, Brussels sprouts, and a few other greens. The need for fresh fruit and vegetables to provide sufficient ascorbic acid in the diet was recognized as far back as the seventeenth century, when it was first discovered that sailors who were at sea for long periods of time often developed scurvy and that scurvy could be traced to a lack of fresh fruit and vegetables in the diet. It was soon established that lemon juice was an excellent remedy for the disease. Ascorbic acid is important for bone formation and bone repair, tooth formation, and especially for wound healing. Long soaking or cooking in large amounts of water is bound to be detrimental to the Vitamin C content of foods.

Vitamin A is also known as axerophthol. It is formed in the body from the yellow pigment of most vegetables, which is known as carotene. The best sources of Vitamin A are asparagus, beet greens, broccoli, carrots, egg, liver, spinach, fish and fish-liver oils, and tomatoes. If there is an excess of Vitamin A in the body it can cause a skin drying, loss of hair, and possibly even fragility of bones. But, far more important, if there is a deficiency of Vitamin A in the body, you will find that your skin will become dry and rough with eruptions, bone and tooth formations will be impaired, and there is a general prone-ness to infection and general debility.

Thiamine is also known as Vitamin Bi. Its best dietary sources are wheat germ, milk and various meats. As in the case of Vitamin C, large amounts of water used in cooking will destroy thiamine, since it is highly soluble in water and can be destroyed by heat. It is essential for normal digestion and is also necessary for the growth and normal functioning of nerve tissue. A deficiency can cause beriberi (common among people whose diet is limited to such foods as polished rice), as well as damage to the heart and to the nerve fibers.

Riboflavin is also known as Vitamin B_2. The best

sources of this vitamin are eggs, kale, liver, milk, and turnip greens. Less rich but important sources of ribo-flavin are kidney, fish, poultry, and green, leafy vegetables such as spinach and mustard greens. Ordinary cooking procedures do not seem to destroy riboflavin in great quantities, though it is water soluble, but riboflavin-high foods should not be cooked in excessive amounts of water. Since milk is also a source of riboflavin, it should be noted that this vitamin can be destroyed by too much light. The use of cartons for milk rather than transparent bottles circumvents this problem and preserves the vitamin which is essential for the healing of wounds and other bodily functions. A deficiency of Vitamin B_2 can cause cataracts in the eyes, dimness of vision; it can also cause scaly skin conditions and lesions around the corners of the mouth.

It should be emphasized that during cooking, foods of high nutritional content may lose some of their nutrients to such an extent that they become poor sources of these nutrients. This happens in one of several ways:

Oxidation. This occurs when foods are exposed to air, and it is hastened by heat or retarded by chilling. The process of oxidation is also hastened by contact with iron and copper, but not by stainless steel. Simmering of foods for too long a time is one way in which the nutrient values can be oxidized away.

Enzymatic action. This is a process which occurs when foods containing agents that hasten and aid in the digestion and metabolism of other foods have this important action destroyed when these raw foods are bruised, crushed, chopped, cut or sliced. For example, cucumber contains an enzyme that destroys Vitamin C in other foods when brought into contact with them. The important enzymatic action of foods is made less valuable by increased heat and is inhibited by cold.

Cooking water. This, if used in too great quantities, will dissolve those vitamins and minerals which do dissolve in water. Cooking water should always be kept to a minimum. Usually a quarter of an inch of water in the bottom of a pan is all that is necessary to cook the vegetables, and the water should not be discarded after the vegetables are cooked. This vegetable water should be utilized in some way—in the making of soup, or cooled and used as a vegetable-juice drink, or used as a sauce, with added salt and pepper, over the vegetable when it is being served. In that way, the vitamins and minerals that do dissolve in water are not being wasted.

Light. This is often destructive to some nutrients, particularly riboflavin (as has already been discussed in the case of milk) and excessive exposure to light will destroy the value of such foods.

The rules for cooking foods are designed, therefore, to minimize exposure to air, light, heat, and such "catalytic agents" as iron and copper. If this is borne in mind while food is being cooked, a great deal can be done to conserve the nutrients of food without having to memorize a long list of detailed rules.

There are some simple rules that can be easily remembered:

1. Fresh vegetables and fruits should be used as soon as possible after they have been purchased. If they must be stored for any length of time, they should be stored in a dark, cool place. Canned vegetables and fruits should be stored in a cool place. Frozen vegetables and fruits should be held at 0 degrees Fahrenheit, with as little variation in temperature as possible.

2. Fresh vegetables and fruits should be handled care fully, since if they are bruised in handling, vitamin losses are high. Paring, mashing, shredding and chopping should be reduced to a minimum. When these have to be resorted to, they should be done just before the vegetable or fruit is served or cooked.

3. If not served immediately, a salad should be refrigerated.

4. Vegetables and fruits should be cooked quickly. They should be added directly to the boiling water and

they should never be overcooked. Some of the original crispness of the vegetable or fruit should be present in the finished product.

5. Soda should never be added to cooking water nor should cooking food be stirred or exposed to light and air any more than absolutely necessary.

6. Vegetables and fruits should not be allowed to stand in water for long periods of time.

7. All waters that have been used to cook fruits and vegetables should not be thrown away. They should be utilized for food, soups, sauces, or in some other way.

8. The quantity of cooking water should be kept to a minimum. As pointed out before, perhaps a quarter of an inch of water in the bottom of a pan is sufficient. The vegetable or fruit does not have to be covered completely by water. Covering the pot will preserve the steam that comes up from the boiling water and will properly cook the vegetable, keeping it crisp, and retaining in the water all the minerals and vitamins.

9. Canned foods and vegetables which have been precooked should be brought quickly to a boil and should not be permitted to continue boiling.

10. Frozen vegetables should be placed directly in boiling water without any prior period of defrosting. An exception should be made for frozen spinach, which should be thawed sufficiently to break the frozen block apart before cooking. Also, frozen peppers may be stuffed and baked either frozen or thawed. (If the vegetables have been frozen in a brine pack, they should be partially thawed before cooking.)

11. Cooked foods should be served immediately and should not be rewarmed or held on a steam table.

12. Short methods of cooking meats, such as pan frying or broiling, are less destructive than the longer methods. Steaming is preferable to boiling, and roasting at low temperature is less destructive to the meat than at high temperature.

Above all, if you choose foods that provide protein at every meal, remembering to serve a green or yellow vegetable every day, as well as to eat one food that is rich in Vitamin C, you are providing for yourself the basis for good nutrition.

* * * * *

Rx ***for Reducing:***

1. **Good nutrition is not faddist nutrition.**
2. **Food preparation should avoid nutritive losses.**
3. **Choose foods that provide highest protein content.**

* * * * *

8: THE HIGH-PROTEIN MENU

If you are overweight, there is only one reason for it. It is because you overeat. Therefore, there is no way for you to lose weight except to eat less. However, eating less does not mean simply eating smaller amounts of the same foods you have always eaten. It does mean that *you have to eat less food of the kinds that put on weight unnecessarily.*

Some doctors firmly believe in calorie-counting as a correct way to diet. However, this is a method that works only for some people. Most dieters have great difficulty in adhering to a calorie-counting diet. On the other hand, most people have little difficulty in adhering to a high-protein diet, simply because the quantity of permissible foods on the diet is not curbed at all. In other words, you may eat all you want of the foods permitted. Under these circumstances, you can never complain of not having had enough to eat. You can fill up.

This diet, as has been pointed out previously, could be called an "exclusively high-protein, low-carbohydrate and no-fat" diet. It permits great quantities of foods that are high in vitamin and nutritive content and generally forbids foods that are low in these qualities.

Foods that are rich in easily available calories are forbidden. The theory is that people who are overweight have a greater number of calories stored in their excess fat than they can possibly use under normal circumstances, and so the weight cannot be lost unless these calories are used up. The principle of the high-protein diet is, then, to utilize the calories that are already stored in the body for the daily energy needs.

The dieter is permitted all lean meats, poultry and sea foods desired. But the visible fat must be trimmed off these foods.

As for fried foods—these are absolutely forbidden.

Vegetables have few restrictions. Only potatoes, corn, peas, rice, and beans of all types except string beans are forbidden. All other vegetables can be eaten in great quantities—cooked or raw.

This diet concept allows for a menu that is extremely healthy and one which can be utilized with countless variations.

While on the diet, the dieter is not permitted any sweets, starches, or fats of any type. And in order to keep to this rule, the diet forbids bread, butter, fat products, cereals, many fruits (there are certain exceptions which will be discussed), and soups (except clear meat bouillon, consommé and other such variations).

Many a potential dieter has immediately complained on seeing this diet. "No fruits?" "Not even a little ice cream?" "Perhaps a small piece of pie?"

There is a one-word answer for all these questions. That answer is: No!

Actually, because the diet is so *simple—and* so *strict-it* is far easier to adhere to it. It is not the kind of diet that permits unconscious cheating. And, as someone who has undoubtedly dieted at least once before—if not countless times—you know the temptation there is to cheat on a diet. This temptation is often unconscious. Sometimes, it is so unconscious that you may not even realize you cheated by having something you shouldn't have had until after you've had it—and sometimes not even at all, though other people will have seen you eating the forbidden foods.

There is no need to worry that your body is not going to get sufficient energy calories to function. These are being supplied by your own body. The fatty tissue that has been stored in your body contains countless calories which were taken from food in which you indulged yourself in the past. These calories are stored in Nature's cleverly economical way, as fat.

Even though the fat intake on this diet is cut sharply, an adequate amount of Vitamins A and D are obtained from eggs, which are permitted at the rate of one a day in the diet, and from yellow vegetables. It should be remembered, too, that even the very leanest of meat has some fat in it. The fact that you trim off every visible piece of fat from a piece of meat doesn't mean to say that you have eliminated all the fat. Good meat is "larded" with bits of fat. This is what makes it tasty. So even on this diet, you are not completely cutting yourself off from the small quantities of fat which the body does need. The fruits and vegetables which the diet permits provide adequate amounts of Vitamin C. And the unlimited quantities of protein foods which are called for in the diet provide a superabundant amount of B-complex vitamins and minerals.

Salt and water are not at all restricted on the diet. In fact, if there is any danger in dieting, it will be *because* salt and water are restricted—unless, of course, for other medical problems your doctor prescribes a salt-free diet. It is true that water is retained in fatty tissue in the body. It is equally true that the salt intake will encourage a larger water-intake. But water does not turn into fat, nor does salt, and once the fatty tissue is worn off your body there will be no place for the water to be retained and the water you drink will not add any permanent poundage nor affect your body measurements.

After having seen what this high-protein diet allows, you may have a thought that you will feel is frightening —that is, that you may have to live on a diet like this for the rest of your life. The answer is, of course, that sacrifices have to be expected in curbing your demands, no matter what kind of a weight-reduction regime you go on. If you go on a calorie-counting diet, you are going to curb your demands. If you go on one of the fancy faddist diets which will be discussed later on, you have to curb your demands. There is no way in which some of your demands will not have to be curbed. However, on the high-protein diet, once your normal weight is attained, you can eat as much as you want—as much as any other normal person—and still hold your weight level. While you are losing several pounds a week, you will have to sacrifice from time to time. When you achieve your normal weight, there will be no need for sacrifice.

The basic diet calls for a small breakfast. Some nutrition experts advocate a large breakfast. It is their belief that if you start with a large breakfast you are starting the day off "right" and you will then be able to adhere to your diet for the rest of the day. It is their contention that a large breakfast enables you to acquire the necessary energy to cope with the day's problems.

However, it has been found that a large breakfast does nothing more than expand the stomach first thing in the day, and thus make the dieter a little more hungry for the next meal. A small breakfast, on the other hand, acts in just the reverse manner. And it has been discovered that most people function quite adequately on, and eventually come to prefer, a small breakfast.

Actually, as one who is overweight, you will probably be able to recall that your eating problem doesn't start in the morning. It usually starts later in the day and in the evening. It is then that you want to snack as well as sit down to a large meal.

The trouble with too many diets for weight reduction is that they have been created on the theory that the dieter is going to fall in simply with the concepts of the expert. This is highly impractical. It is far more practical to consider that the diet should fit with the dieter's actual eating habits. And it was upon this basis that the high-protein diet was conceived.

The basic daily Kotkin Diet* advocated is as follows:

BREAKFAST

One-half small grapefruit
(or one small orange)

*Copyright 1954 by Dr. Leonid Kotkin.

An egg (boiled)
One slice of protein bread
Coffee, tea, or skim milk

LUNCH

Vegetable juice (4 ounces)
Lean meat or pot cheese
Large salad
Coffee, tea, or skim milk

DINNER

Bouillon
Lean meat, fowl, or sea food
One green and one yellow vegetable
One-eighth small melon (in season)
Coffee, tea, or skim milk

With the number of foods permissible, there are a great many palatable variations on the basic meals given. Knowing what you may and may not eat makes it possible to utilize this basic menu in conjunction with a diet menu given by nutritionists. It should be remembered, too, that the basic lunch and dinner can be interchanged as necessary.

The salad suggested for lunch can consist of any of the permissible vegetables—but remember that oil and mayonnaise are prohibited. Vinegar or lemon juice bring out the natural flavor of vegetables, and so do salt and pepper and monosodium glutamate.

Here, once again, should be repeated the basic rules for the dieter. These rules should be studied carefully—and they *must* be obeyed. There are many purposes for each of the rules. If you fail to abide by any one of them, you are going to set yourself the first trap to fall in, and you will be on the route downward to failure.

In order to make your high-protein dieting as easy as possible, the following pages give menus for 21 days with high-protein recipes to go along.

FIRST DAY

BREAKFAST
small glass grapefruit juice
poached egg
one slice protein bread toast*
coffee

LUNCH
carrot and celery sticks
hamburger
wax beans broccoli
tea

DINNER
consommé
broiled swordfish steak
fresh salad with lemon juice
skim milk

SECOND DAY

BREAKFAST
small orange
one slice toast
coffee

LUNCH
small glass vegetable juice
tomato tidbit stuffed with crabmeat (see page 115)
skim milk

DINNER
scallop and radish soup (see page 95)
broiled lamb chop
turnips string beans
tea

THIRD DAY

BREAKFAST
eggs cocotte (see page 99)
one piece toast
coffee

*Reference to toast throughout means protein bread toast.

LUNCH
small piece melon
broiled halibut
spinach boiled carrots
tea

DINNER
tomato juice
broiled calf's liver
mushrooms Brussels sprouts
skim milk

FOURTH DAY

BREAKFAST
half grapefruit
one piece toast
coffee

LUNCH
carrot juice
paprika chicken (see page 100)
green salad
tea

DINNER
bouillon lean
roast beef
asparagus steward tomato
skim milk

FIFTH DAY

BREAKFAST
small orange, sliced
one egg scrambled with skim milk
coffee

LUNCH
tomato and lettuce salad
pepper pot with salmon (see page 115)
skim milk

DINNER
shrimp cocktail
roast leg of lamb
cauliflower frenched green beans
tea

SIXTH DAY

BREAKFAST
tomato juice
one soft-cooked egg
one piece toast
coffee

LUNCH
tuna and vegetable salad
(use brine-pack tuna)
small piece melon
tea

DINNER
tomato bouillon (see page 95)
china cholla (see page 101)
okra carrots
skim milk

SEVENTH DAY

BREAKFAST
orange and grapefruit juice
poached egg
coffee

LUNCH
cold sliced lean tip of tongue
tomato and lettuce salad
skim milk

DINNER
chicken watercress soup (see page 95)
braised lamb (see page 101)
diced beets boiled onions
tea

EIGHTH DAY

BREAKFAST
half grapefruit
one slice toast
coffee

LUNCH
Carrot burgers (see page 116)
tomato and lettuce salad
skim milk

DINNER
tomato madrilène (see page 95)
broiled steak
mushrooms string beans
small piece melon
tea

NINTH DAY

BREAKFAST
eggs cocotte
coffee

LUNCH
mixed vegetable juice
broiled scrod
Brussels sprouts
tea

DINNER
cabbage soup (see page 96)
roast chicken
baked squash beets
skim milk

TENTH DAY

BREAKFAST
orange and grapefruit sections
one slice toast
coffee

LUNCH
bouillon
salmon and vegetable salad
skim milk

DINNER
celery stalks and scallions
roast veal
broccoli mashed carrots (see page 108)
tea

ELEVENTH DAY

BREAKFAST
coddled eggs (see page 99)
one piece toast
coffee

LUNCH
hot borscht (see page 96)
hamburger
sauerkraut
tea

DINNER
vegetable juice
baked fish Creole (see page 106)
small piece of melon
skim milk

TWELFTH DAY

BREAKFAST
half grapefruit
one piece toast
coffee

LUNCH
tomato juice
bell-bottom eggs (see page 100)
skim milk

DINNER
crabmeat cocktail

veal curry (see page 105)
citrus salad
tea

THIRTEENTH DAY

BREAKFAST
small piece melon
soft-cooked egg
coffee

LUNCH
pot cheese and diced green vegetable salad
tea

DINNER
consommé julienne (see page 98)
baked meat loaf (see page 104)
asparagus tips wax beans
skim milk

FOURTEENTH DAY

BREAKFAST
poached egg
one piece toast
coffee

LUNCH
tomato juice
Chinese celery (see page 109)
tea

DINNER
bouillon with thin-sliced radish garnish
baked chicken
tossed green salad
skim milk

FIFTEENTH DAY

BREAKFAST
half grapefruit
one slice toast
coffee

LUNCH
egg broth (see page 96)
tomato tidbit with mushroom (see page 115)
skim milk

DINNER
fennel and carrot sticks
lamburgers (see page 101)
frenched green beans
tea

SIXTEENTH DAY

BREAKFAST
coddled egg
one piece toast
coffee

LUNCH
tomato juice
lean corned beef and cabbage tea

DINNER
beef petite marmite (see page 104)
small piece melon
skim milk

SEVENTEENTH DAY

BREAKFAST
small orange, sliced
one piece toast
coffee

LUNCH
carrot juice
two soft-cooked eggs
sliced tomato
tea

DINNER
baked salmon steak
asparagus vinaigrette (see page 109)

diced
beets skim
milk

EIGHTEENTH DAY

BREAKFAST
small orange, quartered
one piece toast
coffee

LUNCH
citrus sections, pot cheese and lettuce
tea

DINNER
tomato juice
beef à la mode (see page 105)
boiled carrots spinach purée (see page 110)
skim milk

NINETEENTH DAY

BREAKFAST
small piece melon
eggs cocotte
coffee

LUNCH
beet cup salad (see page 110) skim
milk

DINNER
vegetable soup (see page 96)
lamb in foil (see page 102)
small piece melon
tea

TWENTIETH DAY

BREAKFAST
tomato juice
poached egg
coffee

LUNCH
cockie-leekie (see page 97)
skim milk

DINNER
celery
braised tongue (see page 103)
steamed leeks (see page 110)
baked beets (see page 110)
tea

TWENTY-FIRST DAY

BREAKFAST
half grapefruit
soft-cooked egg
coffee

LUNCH
tomato juice
spiced heart (see page 103)
baked carrots (see page 111)
tea

DINNER
watercress consommé (see page 98)
liver Creole (see page 106)
tossed green salad
mashed eggplant (see page 111)
skim milk

RECIPES

(The recipes in the following section include those listed in the daily menus preceding. There are a number of others, which the reader will find good for variety. Your attention is also drawn to Chapter Nine, wherein are found a number of recipes for snacks which can be adapted to luncheon and dinner.)

SOUPS

SCALLOP AND RADISH SOUP

5 scallops
2 pints water
salt and pepper
25 radishes
¼ pint skim milk
chopped chives
paprika

Wash scallops, removing any tough fibers. Chop scallops into small cubes. Peel the radishes. Combine scallops and radishes in a pan with water, salt and pepper. Cover and bring to a boil. Simmer very gently 45-60 minutes. Finally, add skim milk. Warm, but do not boil. Serve sprinkled with chopped chives and a little paprika.

TOMATO BOUILLON

1 can tomato juice
1 bouillon cube
1½ pints water
salt and pepper

Combine all ingredients, bring to a boil and simmer thoroughly for at least 20 minutes.

CHICKEN WATERCRESS SOUP

2 chicken bouillon cubes
2 cups water
1 egg yolk
½ cup finely chopped watercress
salt and pepper

Dissolve bouillon cubes in water and season to taste. Break egg yolk into a bowl and pour boiling bouillon over it, stirring constantly. Serve with chopped watercress on top. The soup can be served without the egg yolk if desired.

TOMATO MADRBLENE

1 envelope (one tablespoon) gelatin
1½ cups beef broth
3 cups tomato juice
¼ teaspoon white pepper

Soften gelatin in ½ cup cold broth. Heat remaining broth and stir into gelatin until dissolved. Add tomato juice and pepper. Taste for seasoning, adding salt if necessary. Chill until jellied (about 4 to 5 hours).

CABBAGE SOUP

1 onion, sliced thin
¼ cup tomato
1 carrot, shredded
¼ cup celery, chopped fine
1 bay leaf
¼ teaspoon caraway seeds
white pepper
salt
½ clove garlic, minced or crushed
consommé
2 cups green cabbage, shredded

Put all ingredients, except the cabbage, into a pan and cover with consommé. Cook 15 minutes. Add cabbage and boil 10 minutes. Serve topped with a tablespoon of yogurt. Try adding ½ teaspoon horseradish and wafer-thin slices of dill pickle.

BORSCHT

clear chicken broth, well seasoned
1 sliced onion (or one chopped scallion)
1 slice lemon, unpeeled
¼ cup celery tops
1 cup coarsely diced cabbage
1 cup coarsely diced raw beets

Blend broth, onion, lemon and celery in electric mixer for about two minutes. Then add cabbage and beets, blending for three or four seconds so that vegetables are barely chopped. Pour into pan and bring to boil. Simmer 5 minutes. Serve topped with a tablespoon of yogurt.

EGG BROTH

bouillon
1 medium egg (per person)

Simmer bouillon in pan. Beat egg lightly, pour slowly into broth and continue to simmer for a few minutes more.

VEGETABLE SOUP

consommé
½ cup parboiled vegetables (per person) (asparagus, string beans, cauliflower, leaf cabbage, celery, green pepper)

Add vegetables to consommé and simmer 5 to 10 minutes.

COCKIE LEEKIE

(This soup is a complete meal in itself)

4 cups water
3 chicken bouillon cubes
2 raw chicken necks, hearts, gizzards (no liver)
4 peppercorns, cracked
4 leeks in one-inch pieces
1 small carrot, diced
2 stalks celery, chopped
pinch of mace
1 turnip, diced
4 thin slices lemon
1 small onion, chopped
salt

Simmer water, bouillon cubes, meat and peppercorns until meat is tender. Remove scum several times while cooking. Cool slightly and skim off all fat. (Hard-to-remove fat can be lifted off by skimming the surface of the liquid with an ice cube held in ice tongs.) Remove the meat and cube it. Meanwhile, add all other ingredients to the liquid and cook until vegetables are tender. Add cubed meat (adding any leftover chicken meat you may have, which should also be cubed).

CUCUMBER SOUP

1 large cucumber
1 small onion, sliced
salt and pepper
2 pints water
2 bouillon cubes
2 egg yolks
1 small jar yogurt
paprika

Peel the cucumber and cut into thin slices into pan. Add the sliced onion, salt, pepper, water and bouillon cubes. Bring mixture to boil, and then simmer gently for about 1 hour, or until the cucumber is quite soft. Season to taste. Now put mixture through a sieve and bring again to boil. Have the egg yolks well beaten in the bowl or tureen it will be served from, and pour the boiling soup over them, stirring all the time. Then stir in the yogurt, mixing until it has a creamy consistency. Serve iced, sprinkled with a little paprika.

If iced soup is not liked, follow the above instructions, but on no account *reboil.* It can be reheated, after the egg yolks and yogurt are mixed in, but it will curdle if it is boiled.

WATERCRESS CONSOMME

1/3 bunch (about í½ ounces) watercress
2 tablespoons chopped parsley
1 large carrot
4 scallion tops
1/2bunch celery leaves
1 cup boiling water
1 cup tomato juice
salt and pepper
parsley for garnish

Wash vegetables. Slice or shred carrots and chop greens. Add vegetables to boiling water and cook covered about 20 minutes, or until carrots are tender. Strain soup and add tomato juice and seasoning. Reheat to boiling point. Serve hot, garnish with parsley.

BEET CONSOMME

juice from 1 can of beets
salt and pepper
2 cans consommé
2 teaspoons wine vinegar

Add the consommé and seasonings to half the beet juice. Heat; add the remaining juice. The object in keeping half the beet juice out is that when it is boiled it loses color.

SWEET PEPPER CONSOMME

3 medium green or red peppers
2 large tomatoes
1 large onion
2 cloves
3 pints water
3 bouillon cubes
salt and pepper

Cut the tops from the peppers and remove the seeds and membrane. Peel the tomatoes and the onion, and chop them all quite finely. Put all ingredients into a pan with the water, bouillon cubes, salt and pepper. Cover and simmer for 11/2-2 hours. Strain, and season again if necessary. Serve hot or cold with a little chopped chives on the top.

GARNISHINGS FOR CONSOMME OR BOUILLON

JULIENNE. Consists of vegetables cut matchstick thin, and about 1-inch long. They should be added about 7 minutes before serving, so that they are cooked but still whole. Chicken can also be added Julienne-style. *POACHED EGGS.* Break the number of eggs required very carefully into the boiling soup and poach for 3 or 4 minutes. Lift out very carefully, put one into each soup bowl, and pour hot soup around it. *ROYAL GARNISH.* 1 whole egg, slightly beaten, 1/4 cup skim milk, 1 egg yolk, salt and pepper. Mix together the whole egg, egg yolk, milk and seasonings. Pour into a shallow baking dish, set in a pan of hot water and bake in a moderate oven for 20 minutes, or until set. Let it get cold, and then dice or cut into fancy shapes. Add to the soup just before serving.

EGGS

EGGS COCOTTE

Pour 2 teaspoons of skim milk into a small casserole. Heat over a slow fire, and then break the egg into it. Add pepper and salt, and let simmer gently on top of the stove. When it is half cooked, put under a moderately hot broiler to finish cooking, bake in the oven for about 15-20 minutes. Tomato juice can be used in place of milk.

CODDLED EGGS

Bring water to a boil, remove from heat. Immerse eggs and let stand 6 to 8 minutes. Break open; mix with cayenne, celery salt, chili powder, chives, curry powder, fennel seed, garlic salt, onion salt, parsley, watercress, or Worcestershire sauce.

POACHED EGGS

Heat salted water to the boiling point in shallow pan. Add one tablespoon of vinegar. Break one egg at a time into a saucer and slide it very carefully into the boiling water. Cover pan, remove from heat. Let stand until eggs are set.

CATALONIAN EGGS

1 large can of tomatoes
1 medium onion
mixed herbs
1 stalk of celery, chopped (optional)
salt and pepper
2 eggs per person

Have a hot oven ready to put the eggs in. Meanwhile place the tomatoes into a pan, grate the onion and herbs into them, add the chopped celery and season well. Simmer gently until the tomato mixture becomes a soft purée. Put into a baking dish and break the eggs into little holes made in the mixture. See that the yolks are unbroken. Put into the hot oven, and bake for about 15 minutes, or until the eggs are cooked but not hard. If you do not wish to use the oven they can be cooked under the broiler, but be careful that the purée does not become too browned.

BELL-BOTTOM EGGS

green peppers
eggs (one for each pepper)
salt and pepper to taste

Select firm peppers that will stand up. Remove stem, seeds and membrane, and parboil 5 minutes. Drain and drop an egg into each, or beat eggs and divide equally among the peppers. Season to taste. Place on baking dish and place in hot oven until eggs are done to desired consistency.

POULTRY

PAPRIKA CHICKEN

1 chicken, sectioned
1 large onion, sliced
2 ounces mushrooms
salt and pepper
parsley, thyme
1 tablespoon paprika
1 pint water
1 jar yogurt

Put the chicken sections in a casserole, add the sliced onion, mushrooms, season well, and sprinkle with paprika. Add water. Cover, and cook in a moderate oven for about 2 hours or longer if the chicken is an old one. Remove from oven and stir in the yogurt just before serving. Serve hot.

MEAT

CHINA CHOLLA

2 pounds shoulder lamb chops
1 large head lettuce
2 packages frozen green beans
3 sprigs of fresh mint
4 spring onions
1 cup consommé
salt and pepper
1 cucumber
2 ounces button mushrooms

Trim the meat of all fat, and either mince or cut it into pieces. Put into a pan, add the lettuce, finely shredded, the green beans, chopped mint, chopped spring onions, consommé, salt and pepper. Cover, and simmer very gently for 12-2 hours, depending on the tenderness of the lamb. One-half hour before serving add the peeled and chopped cucumber and button mushrooms. The mushrooms are not essential, but the cucumber makes the whole dish a delightful fresh summer meal.

BRAISED LAMB WITH CAPERS AND YOGURT

1 lean leg of lamb
1 medium onion, sliced
tarragon, thyme
1 glass water
salt and pepper
2 tablespoons capers
1 jar yogurt

Brown the lamb on all sides in a hot pan, without fat. Transfer to a casserole, add the sliced onion, chopped herbs, water, and seasonings. Cover and cook in a moderate oven for about 2 hours. Fifteen minutes before serving, take it out, and add the capers and yogurt. If you are unable to get an entirely fatless leg of lamb, cook it, and let it get cold, then remove the fat from the top, and reheat.

LAMBURGERS

1 pound ground lamb
1 teaspoon salt
1/4 teaspoon pepper
1/2 cup chopped parsley
1 teaspoon allspice

Combine ingredients and form into four patties. Broil.

LAMB CHOPS FESTIVALE

1 lamb chop per person
2 hard-boiled egg yolks (for4)
2 cloves crushed garlic
1 tablespoon chopped fresh mint
paprika
juice of 1 lemon
yogurt
salt and pepper

Trim the chops of all fat and gristle. Before broiling the chops, have ready the hard-boiled yolks, and mix them thoroughly with the crushed garlic, mint, a pinch of paprika, lemon juice and just enough yogurt to make it a creamy paste. Do not, however, make it sloppy. Sprinkle the chops with salt and pepper and broil them on both sides under a hot flame. Put a spoonful of the sauce Festivale on each chop and serve.

LAMB IN FOIL

1 lamb chop per person (all fat removed)
¼ large onion
½ tomato
garlic salt
½ green pepper,cleaned out
½ small eggplant, unpeeled
powdered basil
salt and pepper

For each serving combine all ingredients, seasoning to taste. Place in a piece of aluminum foil big enough to fold into a roomy envelope. Fold foil several times along edges to retain all steam. Place envelopes on baking sheet. Bake in preheated moderate (350° F.) oven for 1 hour.

SARDINIAN LAMB

1 shoulder of lamb
6 unpeeled garlic cloves
coarse black pepper
6 sprigs rosemary
water
juice of 1/2 lemon

Put the unpeeled garlic cloves underneath the lamb in a baking pan, and cover the top with coarse black pepper. Spread the rosemary thickly over the top. Bake first in a quick oven, then lower the heat slightly and roast for 25 minutes to the pound. Half an hour before it is ready, sprinkle more fresh rosemary on top, pour off all fat, add 1 small glass of water and the lemon juice.

Baste the lamb, but be careful not to disturb the rosemary. When the lamb is cooked, put it on the serving dish, and let the gravy boil up. Add pepper and salt if needed. Add more water if you want more gravy. When eaten cold this Sardinian lamb is delicious.

BRAISED TONGUE

1 pound lamb tongue
water
1 tablespoon vinegar
1 clove garlic, split
1 bay leaf
1 teaspoon mustard seed
10 peppercorns, cracked
1/2 cup onion, chopped
1/4 cup celery, chopped
salt to taste

Cover tongue with water, vinegar, garlic, bay leaf, mustard seed and peppercorns. Cook for one hour. Skin. Cut in slices one-inch thick. Put in casserole with a tight top. Add onion, celery and salt. Add one cup water and bake in preheated medium (350°F.) oven for 1 hour or until tongue is tender.

SPICED HEART

1 1/2 pounds beef heart (calf or lamb heart may be used)
1 onion, minced
1/2 cup shredded celery leaves (or celery stalks)
2 whole cloves
1 small bay leaf (optional)
3 peppercorns
2 teaspoons salt

Be certain hearts are absolutely fresh when purchased. Have butcher remove arteries, veins and all fat. Wash meat well in cold water. If using veal or lamb hearts, cut in halves; if beef heart, cut into 2- or 3-inch slices. Place meat in pan with just enough water to cover. Add onion, celery leaves, cloves, bay leaf and peppercorns. Cover pan and let meat simmer about 30 minutes, then add salt, continuing to simmer until meat is tender (about 45 minutes to an hour for veal and lamb; 1 to 12 hours for beef). Do not let water boil actively. When heart is tender, drain and remove any gristle present. Serve cut into 4-inch slices.

BAKED MEAT LOAF

1 pound ground round beef
1 teaspoon salt
pinch pepper
2 tablespoons minced green pepper
1 tablespoon chopped onion
11/3 cups tomato juice
pinch oregano
1/2 teaspoon liquid artificial sweetener
1/4 teaspoon Worcestershire sauce

Mix well the beef, salt, pepper, green pepper, onion, and 3 cup tomato juice. Pack into baking dish. Bake on lowest shelf of hot (425° F.) oven 10 minutes. Take out and broil 5 minutes. Meanwhile, combine remainder of tomato juice with oregano, sweetener and Worcestershire sauce in a pan and simmer 3 minutes. Remove meat, cut into portions, pour sauce between slices and on top.

BEEF PETITE MARMITE

4 pounds lean brisket of beef
1 pound marrow bones
1 onion
8 medium-size carrots
1 leek
4 stalks celery, diced
2 medium tomatoes, peeled and cut into squares
1 bay leaf
8 black peppercorns
3 sprigs parsley
salt and pepper
1/4 head cabbage

Place beef in heavy pot with bones, cover with cold water. Bring to a quick boil. Remove water and wash brisket. Return to pot. Meanwhile, split onion crosswise and burn cut sides under an open flame. Add onion, carrots, leek, celery, tomatoes, bay leaf, peppercorns, parsley, and salt and pepper to taste, Cover with water simmer 22 to 3 hours. Add cabbage, continue simmering 15 minutes.

To serve, cut brisket in thick slices. Serve with vegetables. This can be served separately as soup and meat: the soup as a broth with the vegetables, the meat with horseradish.

BEEF À LA MODE

Choose an inexpensive lean cut of beef or a rolled roast without the fat. For best results cook four or more pounds at a time. It is excellent cold.

Let the beef marinate overnight in:

1 can bouillon
1 cup water
1/4 cup vinegar
2 large onions, sliced
1 clove garlic
2 bay leaves
3 cloves
1/4 teaspoon oregano
1/4 teaspoon celery seed
10 black peppercorns, cracked
1 teaspoon salt

Brown all sides of the meat on a low flame. Pour off any fat that may have melted off the meat. Add the seasoned marinade and bring to the boiling point. Cover tightly and put into a preheated hot (450° F.) oven. Turn the heat down very low (250° F.) after 10 minutes. The trick is long, slow cooking. A four-pound rolled roast will take 2 to 22 hours. Tougher or larger cuts will take longer.

When the meat is done, put it on a platter and keep it warm. Strain off the liquid into a bowl, discarding all the vegetables. Skim off all the fat. (A good way to get every last globule of fat is to float a clean piece of paper towel on top of the liquid.) Correct the seasoning to taste, if necessary. Slice the meat and pour the gravy over it.

VEAL CURRY

11/2 pounds veal stew meat
1/2 cup chopped onion
1 small orange, peeled
11/2 teaspoons salt
1/8 teaspoon pepper
1 teaspoon curry powder
½ teaspoon ginger
1 artificial sweetening tablet, dissolved
2 cups bouillon (or water)

Brown or sear veal in skillet. Place in large pan, adding all other ingredients. Cover and cook gently until done (about 1 to 14 hours). Lamb will do just as well instead of veal.

LIVER CREOLE

1/4 pound calf s liver (for each person)
1 cup tomatoes
1/2 cup water
1 chicken bouillon cube
1 tablespoon vinegar
1 cup very thinly sliced onion rings
1 clove garlic, minced
1/2 cup green pepper, sliced in rings
1/2 cup mushrooms, sliced through stems and caps
1 bay leaf
pinch of thyme
freshly ground black pepper
salt

Boil all ingredients, except liver, for 10 minutes. Cut liver into 1-inch squares. Add the liver and simmer for 5 minutes.

FISH

BAKED FISH CREOLE

2 cups stewed tomatoes
4 tablespoons thin-sliced onions
4 tablespoons chopped green peppers
4 tablespoons chopped mushrooms
1 tablespoon lemon juice or vinegar
1/2 teaspoon monosodium glutamate
1/4 teaspoon liquid sweetener
pinch of oregano
1/8 teaspoon mustard
2 pounds fish fillets or steaks
salt and pepper

Put all ingredients, except fish, into a pan and simmer, stirring occasionally, until vegetables are tender. Season fish with salt and pepper, place in baking dish, pour sauce over and bake in moderate (350° F.) oven about 30 minutes.

BAKED HALIBUT OR HADDOCK, SWEDISH STYLE

4 halibut or haddock fillets
1 medium onion
1/2 cup parsley
salt and pepper
1 large can tomatoes

Put the fish fillets in a baking dish, slice the onion on top, add half the parsley, season well, and finally pour in the tomatoes. Cover with foil and bake in a hot (425° F.) oven for 30 minutes. Five minutes before it is ready, add the rest of the parsley. Serve from the dish it was cooked in.

TROUT AU BLEU WITH HORSERADISH AND YOGURT SAUCE

4 trout
1 cup wine vinegar
1 cup water
salt and pepper
1 small onion, sliced
1 bay leaf
parsley
thyme

Put the wine vinegar, water, onion, herbs and seasonings into a pan. Bring to a boil and simmer for 1 hour. Strain and heat again. Wrap the cleaned trout in cheesecloth (it is easier to remove them whole this way) and simmer them very gently for about 20 minutes. Drain well, and serve hot with the following sauce:

HORSERADISH AND YOGURT SAUCE

1/2 cup grated horseradish
1 tablespoon tarragon vinegar
juice of 1/2 lemon
1 jar yogurt
salt
cayenne
chopped chives

Combine all ingredients in a double boiler, stir well until the sauce is thick and creamy. Serve either hot or cold.

SPICED FISH CASSEROLE

3 large peeled tomatoes
2 stalks celery, chopped
1 clove garlic, minced
2 ounces mushrooms, dried or fresh
salt
cayenne
nutmeg
1 large onion
1 large sprig chopped parsley
1 teaspoon dried basil
1 bay leaf
2 pounds filleted fish
2 tablespoons vinegar
4 tablespoons water

Peel and slice half the tomatoes and cover the bottom of a casserole with them. Add celery, garlic and 1 ounce of mushrooms. Season well with salt and cayenne and grate a little nutmeg over. Slice the onion very finely and put half on top of this. Cover with chopped parsley, half the basil and the bay leaf. Put the fish on this bed of vegetables, and then put the rest of the mushrooms, tomatoes, onion and herbs on the top. Pour the vinegar and water over. Sprinkle a little more cayenne over, put the lid on and bake in a moderate (350° F.) oven for 30-40 minutes until the fish is cooked.

VEGETABLES

MASHED CARROTS

carrots (as many as required)
paprika
dash of cayenne
dash of powdered allspice
salt

Cook carrots in smallest possible amount of water. Drain and save remaining fluid. Mash carrots through coarse sieve or put them through a ricer. Add seasoning to fluid and pour sauce over serving portions.

CHINESE CELERY

1 bunch celery
salt
2 tablespoons soy sauce
2 tablespoons wine vinegar

Clean and trim the celery, cut it into fairly thick strips, and bring it to a boil in salted water. Let it boil for 5 minutes, no longer. Drain, and while it is still hot, add the vinegar and soy sauce. It can be eaten hot or cold, and will keep for some days if the celery is covered by the sauce. It is also a very simple, yet delicious hors d'oeuvre.

Fennel, turnips or radishes can be prepared in the same fashion.

ASPARAGUS VINAIGRETTE

Either fresh, frozen or canned asparagus can be used. For fresh or frozen asparagus, cook in little bundles in boiling salted water for about 10-15 minutes, depending on the size of the heads. Do not overcook it. Test with a fork in the stems. Drain carefully, and leave to get cold. Reserve a little of the water it was cooked in for the sauce:

SAUCE VINAIGRETTE

1 hard-boiled egg
2 tablespoons tarragon vinegar
1 tablespoon capers, cooked chopped
2 gherkins, chopped
salt and pepper
6 tablespoons of the strained liquid the asparagus was in

Remove the yolk from the hard-boiled egg and mash it thoroughly. Gradually stir in the vinegar, and then add the chopped capers and gherkins, salt and pepper. Now, mix in the asparagus liquid, stirring until smooth. Add finely mashed egg white. This sauce can be served hot or cold.

SPINACH PURÉE

spinach (as much as required)
dash of nutmeg
pepper
salt
Worcestershire sauce

Wash the leaves very thoroughly. Cut off the tough stems and ribs unless spinach is very young. Put the leaves through a grinder and then into the top part of a double boiler with the nutmeg, pepper and salt. Cook 30 minutes, stirring occasionally. Just before serving, stir in Worcestershire sauce (½ teaspoon to each cup of puree). This purée can also be served as a base for grilled tomato or poached egg.

BEET CUP SALAD

beets (as many as required)
1 teaspoon minced onion
1 teaspoon horseradish
1 tablespoon chopped cucumber

Boil the beets in heavily salted water. Chill. Hollow them to form fairly thin cups. Dice the centers and add onion, horseradish and cucumber in the amounts indicated for each serving. Heap into the beet shells and serve on a lettuce leaf.

BAKED BEETS

beets (as many as required)
1 cup strained orange juice
dash of cayenne
1/4 teaspoon salt

Select young, tender beets. Peel and slice thinly. Place in baking dish. Add juice and spices. Bake, covered tightly, in a preheated hot (400° F.) oven for about 30 minutes or until tender.

STEAMED LEEKS

Remove most of the green top, leaving about four inches of the firm green part above the white portion. If large, split lengthwise. Wash thoroughly. Lay the leeks on a rack a few inches above boiling water. Cover tightly and steam for 20 minutes or until leeks are tender, depending on their size. Serve with Sauce Vinaigrette (see page 109) or with salt, pepper and lemon juice.

BAKED CARROTS

young carrots (as many as required)
1 cup water
1 chicken bouillon cube
1 small onion, grated
1 teaspoon Worcestershire sauce

Arrange carrots on bottom of a baking dish. Add water, bouillon cube, onion and Worcestershire sauce. Bring to a boil. Cover and bake in a preheated moderate (350° F.) oven for 30 minutes, or until tender.

MASHED EGGPLANT

eggplant
white pepper
onion salt
paprika

Peel and cut an eggplant into pieces. Cover with boiling salted water. Simmer until tender. Drain water (this should be saved for a soup dish, adding a bouillon cube to each cup of liquid). Beat until fluffy. Season with pepper and onion salt. Dust with paprika before serving.

* * * * *

Rx *for Reducing:*

1. **You must cut down on fattening foods.**
2. **You must stick strictly to the high-protein rules.**
3. **If you fail to abide by the rules, you defeat your diet immediately.**

* * * * *

9: SNACK AND STAY SLIM

The last time you tried to diet, did you succeed?

The answer is most probably no; otherwise you wouldn't be reading this book and seeking a new way of losing weight. Have you any idea why you didn't succeed? The chances are one of the main reasons was that you cheated. You cheated yourself—let's make that clear. You weren't cheating anyone else.

Most people fail at a diet regime because they cheat. In fact, they make no bones about it. Most of them admit it quite readily. One person will say: "I just can't help taking a little snack before bed." Another will say: "But I get so hungry in the afternoon, I have to have a bite to eat."

Virtually every unsuccessful dieter you talk to has a similar answer. And for every answer that spelled doom to diet, there was an accompanying excuse. Each excuse, the dieter felt, was perfectly valid. In the language of the psychiatrist, these explanations for diet-cheating are only rationalizations.

Excuses or rationalizations; the fancy language doesn't change the fact that cheating is cheating. It leads inevitably to the downfall of the well-intended dieter's carefully laid plan to reduce.

This little expose of cheating shouldn't discourage you, however. In fact, under the high-protein diet, as you may have already guessed, you can cheat.

"All the snacks I want?"-"I just can't believe it!"

These thoughts are undoubtedly going through your mind. As a previously Determined Dieter whose determination lagged somewhere along the way, this undoubtedly sounds a little bit incredible to you. But there is nothing incredible about it. Other people-countless thousands—have asked the same question and made the same exclamation. These statements were made before they tried the high-protein diet. After they tried the diet —usually with great success—the comments went something like this: "Just amazing; I ate and ate and ate. Later I no longer had the appetite to gorge myself so much. But despite the eating, I still lost weight."

"During the early days of the diet, I actually cheated quite often—and best of all, I didn't feel guilty about it. I lost twenty-eight pounds in less than two months/'

"Since I understand why I was putting on weight, I've had no trouble keeping the scale under control—snacks or no snacks."

These were the comments of three skeptics who turned into three svelte disciples of the high-protein diet.

So, as a snack-sneaker, you needn't lose heart because you have failed before at dieting. The simple truth is that with the high-protein diet, you have the key to success: you can sneak those snacks and still stay slim.

Why does a dieter cheat? Usually for one of two reasons: either he feels that the diet he is on is starving him, or he unconsciously has a will to fail at a diet regime. The first reason is one which stems from poorly conceived diets.

There are diets which starve a person. These appear in the public press almost every day. Many are created by people who are not in the least qualified to act as nutritionists or to prescribe as physicians. Yet they willy-nilly set about creating some faddist diet and easily encourage thousands to try it.

The results can be disastrous. Fortunately, practically everybody who tries such a diet gives in when he finds that he lacks sufficient energy to do even simple daily tasks.

Other diets, which are designed by actual nutritional experts, fail to take into consideration the "personal factor." Everybody has different needs. These needs are dependent upon a person's individual physical and organic makeup. They are also tempered by what work the person does, by what hours he keeps, by a host of other personal factors. Therefore, a diet has to have basic design and must be flexible enough to fit everyone's own individual needs.

The high-protein diet—giving you a wide variety of foods with only the restriction that they *be* high-protein foods-fills this bill ideally.

What about the will to fail at a diet regime? This is a reason which many doctors have now realized is the greatest temptation to the diet-cheat. It is created in the same unhealthy pattern which, in the first place, led the overweight person to become too heavy. In other words, medicine has come to realize recently that there are emotional causes that create a desire to eat. The idea that overweight was caused by some organic or metabolic malfunction of the body has been completely disproved. There is no truth to the theory that glands cause overweight. The only thing that glands do in connection with overweight is to distribute the excess fat to various parts of the body, depending upon which gland is involved and how it may be malfunctioning.

If there is a desire to place the blame for overeating on a gland, the only gland that should be blamed is the salivary gland. In other words, to reiterate, people are overweight because they overeat.

So, recognizing the fact that emotional causes set up a desire to eat in the first place, you must now recognize that these same emotional disturbances will encourage the will to fail at a diet. This failure drive will lead the dieter to unconscious cheating-cheating, which after a short time will show markedly on the bathroom scale and force the discouraged dieter to give up in disgust.

"I just can't manage to diet," is an excuse frequently given by overweight people. "I stick to my diet as closely as possible during the three meals a day—but it's between meals that the trouble begins."

"I frequently get hungry in mid-afternoon or in the evening. Before I know what I'm doing, I've had a snack." Well, at last, you have a "magic" answer. You can go ahead and have a snack.

If you're hungry between meals, *eat.* However, *eat foods that are on the basic high-protein diet.* Limit yourself to fresh vegetables, to lean meats, to poultry, to fish, or to pot cheese. And just as you have already discovered that the limitations of the high-protein diet make for an appetizing and multi-variety menu for your basic three meals, so you will be surprised at the tasty, and filling, snacks that can be whipped up for your between-meal hungers. For instance:

PEPPER POTS: Crisp *green peppers,* halved, cleaned, stuffed with:

- *crab meaty mashed tomato* and *cubed cucumber, lemon juice* and *salt* to taste.
- *salmon* and *cucumber* (cook pared cucumber for 10 minutes, or until soft, in small amount of salted water, mash and season with salt and pepper, mix with drained salmon), season with lemon juice to taste.

TOMATO TIDBITS: Large *tomato,* halved or quartered, emptied of seeds, pulp and most of liquid, sprinkled with salt and inverted for 30 minutes. Save pulp to add to stuffings, such as:

- *mushroom,* chopped finely, mixed with tomato pulp, seasoned to taste.
- *crab meat* added to about ½ tablespoon of chopped *red* or *green pepper.*
- *egg,* hard-boiled, yolks mashed with tomato pulp and finely chopped *onion.* Garnish tomato sections with whites after putting them through potato ricer.

OPEN-FACED SANDWICHES: (one slice of protein bread)

- *pot cheese* with chopped ripe *olive,* salt to taste.
- *tuna fish* (flaked and dried on paper towel) mixed with chopped *chive.*

- *chicken salad* (chicken meat, *celery* and *carrot* put through meat grinder, mixed with *tomato* pulp to moisten), lemon juice and salt to taste.

Easy things to keep in the refrigerator:

Small *tomato* and *lettuce* salad, lemon juice, salt and pepper to taste. For different taste, add diced *green pepper, shallots, celery* and *cabbage* heart.

Or keep a dish of cold water, with a touch of lemon juice added, in the refrigerator, filled with tender *celery* hearts and *carrot* sticks for a quick snack. Season with salt before eating.

A lean *hamburger* patty and ketchup (season with a little finely chopped *onion* and/or *green pepper* before grilling, salt and pepper), serve as an open-faced sandwich on one slice of protein bread or as a filling for a Pepper Pot or Tomato Tidbit.

If you're entertaining, and want to go to a little more trouble, why not try some of these:

CARROT CANAPÉS: Scrape large even *carrot.* Cut off ends and make hole through center with apple corer. Stuff tightly with highly seasoned *pot cheese* mixed with finely cut *chives.* Chill until cheese is firm, cut to 4-inch slices and serve.

CARROTBURGERS: Prepare *carrots* as above, stuff with lean *hamburger,* mixed with finely chopped *onion* and raw *egg.* Bake in medium oven until carrot is tender. Cut into 2-inch slices and serve.

CUCUMBER CUPS: Wash *cucumbers,* cut in halves crosswise, remove seeds and fill with *chicken* which has been put through a meat grinder and mixed with a few grains of *cayenne* and a slight grating of *nutmeg.* Add *skim milk* to make paste, stuff cucumbers, place upright on trivet in a saucepan and add stock to half-cover cucumbers. Cook covered, 40 minutes. Serve hot.

CUCUMBER CUTUPS: Wash cuke, stripe skin with a silver fork. Slice to 1/4-inch thicknesses, then cut three-quarters of the way into each slice and insert thin slice of *radish.* Salt to taste and serve ice cold.

CELERY BOATS: Wash and dry pieces of *celery* from the heart. Leave a bit of foliage, stuff with chopped raw *mushrooms* seasoned with *Worcestershire sauce* and *garlic.*

Of course, there are other simple and easily prepared snacks that you can keep in your refrigerator. For example, a raw cauliflower should be taken apart and the individual buds stored in the refrigerator. They are an extremely tasty snack, especially taken with salt or lemon juice. Other vegetables such as cucumber, carrot, radish, green pepper, tomato, celery, iceberg lettuce cut into firm quarters, fennel—all of these make fine quick-to-prepare snacks that will take the place of your usual snack of a piece of pie, a slice of cake, a dish of ice cream, or what have you.

Since you have to steer clear of all liquor—beer, for instance, contains 86 calories for every six ounces, while a Martini has 143 calories—there are any number of low-calorie soft drinks on the market, but why not iced coffee with artificial sweetening, or iced tea with artificial sweetening and some lemon?

Bouillon or consommé on the rocks makes an ideal cold drink—not only refreshing but quite filling and fitting ideally into the diet. A glass of vegetable juice-season it all you like—or a small glass of skim milk will certainly serve to quench your thirst.

As for salad dressings, lemon juice is an ideal dressing for most salads; so is a touch of vinegar. But you'll find that your salads will taste most refreshing with salt or monosodium glutamate (popularly packaged under the trade name of Accent).

For a sharp-tasting snack, once in a while, why not a dill pickle?

* * * * *

Rx *for Reducing:*

1. **Snacks need not spoil your diet.**
2. **Between-meal snacks must follow the high-protein diet.**

* * * * *

Part Three: STICK TO IT!

10: FATS, FADS AND FALLACIES

It was Ben Jonson who once said, "Better be dumb than superstitious." But, despite the poet's admonition, despite the high level of education in our world today, superstition runs rampant.

Are you superstitious? Do you believe that if you break a mirror you'll have seven years' bad luck? Or that if you walk under a ladder, ill fortune will befall you?

These two superstitions may be beneath you, but maybe you find that you "touch wood" every time you want to assure yourself of preserving good fortune. Actually, commonplace items and occurrences are involved in hundreds of common superstitions, and there is hardly a person alive who doesn't, in some way, hold some superstitious belief.

Unfortunately, probably the widest held group of superstitious beliefs is in the field of food faddery. Whether he is savage or civilized, man eats for a number of reasons. Among these are the appeasement of his hunger, the satisfaction of his appetite, force of habit, sense of duty, a means of promoting social intercourse, an alternative to the more direct but unobtainable or socially unacceptable satisfaction of other urges, and as a device for winning recognition by his fellows.

Man may refuse to eat certain foods, or to eat at all, because to do so would be to conflict with one or more of his cherished beliefs or ambitions, or as a gesture of defiance to society or toward other people with whom he is in emotional conflict.

Due to man's intellectual superiority over the lower

animals, his greater powers of imagination, reasoning, and means of exchange of ideas and experiences with his fellows, he has succeeded in unduly complicating his own feeding and nutrition problems—while doing much to improve the nutritional welfare of the animals which he has domesticated! It seems that man has been able to think of more reasons why he should not eat foods of good nutritional value than would be possible for an animal of feebler intellect, or for one with more knowledge and better judgment.

If you were to collect the list of beliefs concerning food and diet that appeared in newspapers and magazines in the past two or three years alone, you would have a list so long that it would virtually fill this entire book. And yet, when you read some of these beliefs and taboos about food, you are going to see how highly irrational they are and how implausible it is that they could have developed.

For example, here are some popular beliefs about foods and nutrition which are generally recognized to be *false:*

Diets high in red meats cause high blood pressure and kidney disease.

Red meat in a diet makes one vicious.

A mother should not eat meat after giving birth.

Meat and dairy products should not be eaten together or at the same meal.

The combination of ice cream or milk with shellfish is poisonous or indigestible.

Pickles and ice cream are an indigestible combination.

Fish is "brain food."

Butter is rich in protein, and is more nutritious than margarine.

The top of the milk is best for children.

Sour cream is less fattening than sweet cream.

Apple sauce should be served with greasy foods because it absorbs the grease and promotes elimination of excess grease.

No breakfast is complete without cooked or prepared cereal.

Black coffee is "stronger'' than coffee served with milk or cream.

One should not eat before going to bed.

Three meals a day are essential for good health.

Brown eggs are more nutritious than white eggs, or *vice versa.*

An apple a day keeps the doctor away.

Beets make the blood rich.

Lemon makes the blood thin.

Tomatoes and citrus fruits make the blood acid.

Sugar in the diet will cause diabetes.

A fat man is a happy man.

This is just a short list. For the most part, it doesn't deal with the fads and fallacies with which we are more directly concerned at the moment—those revolving about dieting.

The superstitious beliefs surrounding the problem of overweight have become a serious block to countless thousands who would otherwise be able to lose weight successfully. It seems that these people become discouraged at hearing one or other of the stories—all of which are, in the main, false—which concern the merits of obesity, the perils of dieting, and the vast body of misinformation about fattiness.

Let us look at some of these and see if we can't dispel them quickly. You may be guilty of believing any one or more of them. That would not be surprising nor would it put you in a class by yourself. If you do honestly believe in any of the superstitions that are listed above, or below, you have much company. However, you need not be proud of the fact that you are not alone. Every one of these superstitious beliefs has been incontrovertably disproved, thanks to science.

It's healthy to be fat. The very opposite is true: it is unhealthy to be fat. The Metropolitan Life Insurance Company has compiled statistics over the years which show definitely that overweight persons have a far greater death rate than do persons whose weight is normal or even below normal. Statistics show that among those who are overweight there is a greater amount of high blood pressure, hardening of the arteries, gall bladder diseases and diabetes. Overweight persons are also poorer risks on the operating table and they tend, as well, to recover with greater difficulty from, or to succumb with more ease to, serious illnesses.

You can die from dieting. This is a belief which is, surprisingly enough, held quite seriously by too many people. It is based mostly on stories told by relatives of persons who have succumbed to "wasting diseases." And once the seed of this belief has been planted, a person is very apt to find himself associating the wasting effects of tuberculosis and cancer with dieting and the resultant loss of weight This is indeed unfortunate, since dieting to lose excess fat is not starvation. Yet many stories circulate about people who actually have died of cancer, or some other wasting disease, to the effect that they were on a diet and that is why they had lost so much weight. This is often the case because people tend to regard with shame cancer and other diseases of a wasting nature. They would rather have their friends and acquaintances think that a diet was responsible for the loss of weight and the resultant death. Of course, when the so-called dieter dies as the logical outcome of the fatal disease, the legend is perpetuated because of mistaken shame. It should be firmly stated that no one has ever become seriously ill through dieting to lose excess fat, when the dieting was done properly and, especially, under the supervision of a physician.

Body resistance is lowered while dieting. This is a widely accepted piece of folklore among people who believe that they are more susceptible to colds and other similar infections every time they go on a diet. It is odd that they don't recognize that when they have a cold or a virus infection, many of their friends around them also are suffering in the same way. But, because they are on a diet, they tend to couple the two factors and insist that their body resistance was lowered and that, therefore, they have become ill. Sometimes the dieter will not blame the diet, but friends and relatives will be quick to point out that the infection probably was aided by the fact that the dieter was not getting "proper nutritive intake." It should be re-emphasized once again that just as long as the vitamin, mineral and protein intake is adequate, the body will not become deficient in any element necessary for good health.

It's less dangerous to diet slowly. The answer to a statement of that sort is that neither is it dangerous to diet nor is it dangerous to diet quickly. Dieting quickly does not mean starvation. A weight-control program which includes protein foods, vegetables and supplemental vitamins will not permit starvation even though fat may be lost from the body rapidly. People who prefer to diet slowly often use this only as an unconscious excuse to avoid giving up the satisfactions of eating all the rich and sweet foods that they love so well. A person who is sixty pounds overweight, for example, may lose seven pounds during the first week of a sincere weight-reduction effort. This is mostly water which stores itself in the fat. Perhaps only twenty-five per cent of the total weight loss is actually fat, and after a while the same dieting effort will only yield about two pounds a week. Eventually the loss will drop to about one pound a week. The amount of fat lost during a weight-reduction program is practically the same week by week. However, if the dieter fails to maintain his pace of diet, the change in weight loss will be proportionate. Often a patient who succeeds in accomplishing a large loss in the first week will lessen the effort in the flush of success. In effect, however, due to the bulk of water loss in the initial week, the diet program will come to a standstill long before the desired weight is reached.

If weight is lost slowly, normal weight can be maintained longer. Experience has not borne this out. Patients who diet slowly, with difficulty and with complaints of deprivation, will regain more rapidly the weight they lost and also will have far more difficulty in maintaining the weight they want. On the other hand, patients who "take hold" of themselves, who throw themselves into a diet program and lose weight rapidly have, in general, shown themselves to be good "weight-watchers" and many of them never regain the weight they took off.

Overweight prevents nervousness. There may be some superficial truth in this, since some people claim to feel "better" when they overeat. However, this is only because the overweight person is cloaking many of his disappointments in oral satisfaction. This is an indication of a rather serious emotional problem and medicine has come to the belief that it is far better for such people to reach a state of normal weight and perhaps suffer some small nervousness because this is the only way in which they have a chance of facing reality and of eventually solving their problems in a mature manner.

Fat people are happy and jolly. And the reverse is often heard: *thin people are ill-tempered and unhappy.* There is some truth in the belief that fat people appear jolly. But they do so to mask their true feelings. The reputation of the fat person for jolliness is really undeserved. People who are excessively overweight are usually insecure and feel inferior. Therefore they are almost unconsciously motivated to be agreeable to their relatives and associates and friends. They feel that the only way they can win love and admiration and respect is by being good-natured and by allowing themselves to become doormats for anyone who wants to step on them. This superstition gets further commerce when the obese person proves to be a successful weight-reducer. It seems that when such a person attains normal physique, he will instinctively show a new willingness to fight for his rights. This, of course, will be used by the superstitious to reinforce their belief that the fat person is a happy and jolly person. They will point to the successful dieter and say: "See how disagreeable he has become now that he has lost weight?"

Fat runs in the family. The tendency to overweight is definitely not inherited. Physique, body build and stature can be inherited—but not fat. In families where obesity tends to be prevalent, the only reason for this is overemphasis on eating habits. These are usually families in which the daily menu comprises rich, starchy and fried foods—served often and served in large quantities. But while one's body build may be inherited, the fat that cloaks those bodies can only be obtained in one way-overeating.

Overweight is a glandular problem. Medicine has clearly determined that while certain glands in the body may determine the kind of fat that the body has and the distribution of that fat around the body, glands themselves do not create fat. Especially, they do not create fat in any mysterious fashion out of thin air, as is implied by overweight people so often. If, indeed, there is any gland in the body that can be responsible it is—as has been pointed out before—the salivary gland. Your desire to taste rich foods is the only desire that leads you into overeating, and overeating is the only way in which you are led into overweight.

Exercise is the best way to lose weight. This is a great trick if you can do it. Unfortunately, while exercise will tighten up the muscles of the body, it will not succeed in losing weight for you. It will, in fact, increase your appetite from time to time. Exercise and its value, however, will be discussed further in the next chapter.

Water turns into fat. This is really a "lulu" of a belief. It belongs in the same category as the belief that certain glands in the body will turn air into fat. Here, too, is a good trick if you can accomplish it. The body will store excess water temporarily during a weight-reduction program and this will result in a "lag period," when weight will hold fairly steady and it will seem as though no fat is coming off the body despite the dieter's great effort. Actually, however, fat is being lost from the body during this period even though the body retains water for a while, or so it seems. During a diet regime, your weight equals the normal you, plus the excess fat that you are carrying, plus or minus water. The water factor in the equation is unimportant, since it is not held by the body for any length of time. The body can only hold a limited amount of water, in any case. When the water is held, a lag seems to take place in the reduction program. However, it will only result in a greater drop in weight later on while you are dieting. The idea that water will turn into fat is, however, completely baseless. There is absolutely no mechanism in the body—or, for that matter, out of the body—that is capable of making fat out of water.

The stomach shrinks permanently on a diet. Many a dieter has found that after a long period of eating less food he no longer has the desire for, or feels the need of, as much food as he did. Thus, he becomes more easily satisfied with amounts of food that would previously have left him greatly unsatisfied. Many dieters attribute this feeling to a "shrinkage" of the stomach—a process which does not occur and cannot occur. The stomach is only a muscular elastic organ. It automatically accommodates itself to vast changes in the amount of food and drink placed in it. What probably does occur when a dieter feels his stomach has "shrunk" is that there has been an actual change of attitude toward food and the dieter has realized he is able to function more efficiently and stay in better health when he is taking reduced quantities of food.

Better absorption of food nourishment causes fat. In other words, some people will tell you, thin people waste their food and fat people get all the good benefits from it. Science has been able to disprove this completely simply by carefully examining the excrement of both thin and fat people. This kind of a survey showed that the amount of calories excreted by both types of persons varied so little as to be of no importance whatsoever.

Toast is less fattening than bread. A slice of bread, whether toasted or plain, contains exactly the same food composition, and "burning" the bread does not burn away any calories.

Wheat and rye breads are less fattening. Some "dark" breads contain more malt than white bread and are therefore more fattening.

Meal-skipping is a good way to reduce. As often as not, going without one meal will make you eat more at the following meal. It is better to eat three balanced meals a day in order to reduce.

Massage will make you slim. Massage has no material value in weight reducing. Its function is to stir up lazy muscles and tissues which allow excess fat to sit on them.

Surgical operations cause people to gain a lot of weight The operation causes no weight increase, but the inactivity necessary after an operation and the food that is often prescribed in convalescence will often cause a gain in weight.

It is natural to gain weight as one gets older. Your weight should not increase more than ten pounds after the age of twenty-five. Life insurance companies give far more life expectancy for thin people over the age of forty.

Salt is fattening. This falls into the same category as the "water is fattening" theory. Salt is not a food and therefore it is not fattening.

Gin makes you thin. This is a rather interesting one which appears to have come to the United States from some of the European countries where gin drinking is more prevalent. In fact, quite the opposite is true. Gin, like all liquors, has a high food value. Instead of making you thin, it can have quite the opposite effect.

Another area in which fads have taken hold is in the area of diets themselves. There are probably more "diets" being passed around from person to person—without the consultation of a doctor in any way—than anything else except advice or money. The only difficulty is that often a diet prescribed by a particular doctor for a particular patient in a particular instance is useless to anyone else under any normal set of circumstances. But, worse than being useless, many of these diets can be dangerous because of differing circumstances.

As you have already seen, there is nothing simple about nutrition. Brilliant men spend their lives researching it. On the other hand, there are some basic rules which anyone can understand, rules which medicine has discovered about "improper" eating. Self-prescribed and extreme reducing plans are so prevalent that they are usually dangerous.

Raw-vegetable-and-fruit diet. Any doctor will assure you that such a diet is an express route to serious trouble. Yet thousands of people prescribe it for themselves. They have the mistaken idea that this is a health diet as well as a reducing diet, that it "cleanses the system" as it takes off pounds. It does not In fact, it is much more likely to produce either constipation or diarrhea and other serious ill effects. This kind of diet should not be confused with a so-called vegetarian diet. The vegetarian, if he remains in good health, adds quantities of milk, eggs and cheese to his ration of vegetables, fruits, cereals and nuts. So he gets enough of the animal proteins which are essential to everyone. If you try to lose weight by eating nothing but vegetables and fruits, nature will give you a sharp warning that it will not tolerate such a lack of protein—protein that is best supplied by lean meat, poultry, fish, eggs and milk. Within a day or two you will feel weak and the weakness will not pass no matter how much you stuff yourself with great quantities of carrots and celery and apples. You will lose weight, but not necessarily much fat. The lost weight will be stolen from your muscle tissue as well as your reserves of fat and you will rapidly put yourself into low gear. Your skin will become dry and scaly. You will find yourself jumpy and cross and developing nagging headaches and a listless inability to concentrate on anything.

Banana-and-skim-milk diet. Unfortunately, this is another diet that too many people often prescribe for themselves. There is no doubt that anyone who eats nothing but bananas and milk and a little unsalted lettuce or spinach for several weeks will lose a considerable number of pounds. But if a diet of this sort is undertaken without a doctor's orders, the dieter is going to find himself suffering from more than just overweight. He will lose health as quickly as he loses weight and will probably put back most of the lost weight as soon as he discontinues the diet and goes back to his old eating habits. This is not only a useless diet, it is a dangerous one. It should be emphasized that bananas and milk are both superb foods and contain a goodly assortment of the elements we need for health. It is, however, the deficiencies that such a diet has—as, for example, iron—which cause the trouble. This diet will lead to anemia—a miserable, stubborn malady from which recovery is usually a long struggle.

Vegetable-fruit-and-milk diet. These are all wonderful foods—but, by themselves, they are inadequate. You can stay on this diet a little longer than you can stay on the raw-vegetable diet because you won't feel quite so bad as quickly. However, you will soon find that your skin will erupt, your digestion and elimination will prove troublesome, and a general weakness from protein deficiency and a lack of niacin, which comes mainly from lean meats, will soon appear. A lack of niacin in the diet will cause pellagra, a disease that was prevalent in the southern United States until it was discovered that the diet of those people who contracted it consisted mainly of salt pork, corn bread, and molasses. Of course, you aren't going to develop pellagra if you go on such a diet for a short time. But if you cut out meat, eggs, and cheese completely, you will surely lose vitality and stamina.

Grapefruit-and-black-coffee diet. The basic plan for this strange diet calls for nothing but grapefruit and black coffee at breakfast and lunch. At dinner the dieter is allowed one helping of lean steak, a baked potato without butter—and more grapefruit and coffee. Those are certainly healthful foods, but taken alone they are going to make you ill long before they make you slim. You will find yourself very susceptible to any cold or influenza germs or any virus which is floating around. This diet is particularly defeating for older people, whose skin is less elastic, and they quickly get a wrinkled look.

Salt-free, rice diet. A doctor may prescribe this as a weight-reducing plan for a particular reason. But anyone else who goes on this diet without the specific prescription of a doctor is clearly out of his mind. Nothing but grave trouble can result from it unless it is taken under doctor's orders under a doctor's strict supervision. It is an extreme measure used by doctors to fight certain serious troubles. It is usually administered under clinical conditions—that is, in a hospital so that it can be completely controlled. Leave this one strictly alone!

High-fat diet. The latest fad diet—that of eating highly fatty food in great quantities—is one which some medical authorities have come to think of as the weirdest of all diets. It actually is not as new as some would have you think. It was first developed about a dozen years ago and has been discounted several times in the interim. But a recent best-selling book has placed this diet onto the news pages once again—with some disastrous results. The American Medical Association, for instance, feels that it "is a grave injustice to the intelligent public and can only result in considerable damage to the prestige of the medical profession." And a world-famed nutritionist said of the diet's propounder that his training "does not necessarily imply any competence in nutrition and biochemistry." It has been discovered that a high-fat diet may reduce weight temporarily through dehydration of body tissue—but it cannot work permanently because of basic medical law which states that a person gains weight if he eats more calories than he burns up. And on a high-fat diet, the caloric intake is so high it is virtually impossible to burn away. Part of this theory has become involved with polyunsaturated fats, which everyone has become conscious of because of the part they play in heart disease. But fats are fats, and excess fats are going to be stored as fat in the body no matter what kind they are or what their source is. In addition, nutritionists point out that this high-fat diet will in the long run be harmful, since it is less likely to provide sufficient vitamins (especially A and C) and minerals (especially calcium). The diet will activate gall bladder troubles and other troubles as yet unknown. And persons with diabetes or kidney and liver disease might be harmed.

Crash diets. There are several hundred—and new ones being added almost daily—crash diets that virtually every overweight person tries at some time or another. Some of the more unusual ones have been a *watermelon-and-grape* diet, a *boiled-egg-and-skim-milk* diet, a *mushroom-and-celery* diet—and even one that involved dry *Martinis!!* The latest diet enthusiasm is for a flavored, liquid, 900-calorie formula based on milk products, soy bean flour, and vegetable oil. The meal is over in one long drink. The formula is an improvement over most of the other crash-diet programs simply because it is so convenient. It can be taken along for your lunch on the job, and for the woman at home it offers an artificial prop simply because she does not have to handle food or prepare it in any way—and therefore there is no added temptation. Since the introduction in September, 1959, of Metrecal, the first of the current formula products, there has been a succession of similar preparations marketed in a wide price range—from less than fifty cents to more than three times that amount for a daily supply. Their commercial success has been dazzling. Because formula diets of this sort do depart from the standard dietary principles emphasized earlier in this book, and because they are so popular with the public at the moment, the following facts about them should be known:

1. Used as directed, there will undoubtedly be some weight loss. But the amount will vary from person to person. Part of the immediate loss is a water loss as in most diets because these formula diets have a low salt content.

2. The long-term effectiveness of these diets—that is, the maintenance of weight loss after the diet ends-appears to be comparable to that of other reducing pro grams which are unsuccessful on a long-term basis for the overwhelming majority of dieters.

3. It is probable that many people will not be able to stay on this kind of diet for long periods of time.

4. Used for *short* periods without medical supervision, these diets don't seem to pose any health hazard that is

too serious for most people. However, since experience indicates that these products frequently are used as a substitute for one or two meals a day, it has been found that the regular food intake at other meals is far greater than it should be and is often a counter-balance to the effectiveness of such a diet.

5. This must be noted well: There are some people with specific—and possibly undiagnosed—ailments who can be seriously hurt by staying on a crash diet of any sort (including these liquid diets) for any long period without medical advice.

6. These formula diets, in themselves, are not a miracle cure for overweight. They have no magic properties that will help you lose weight easily, and permanently. Used as an occasional feeding to replace some regular meals, they can be partially effective and they are certainly harmless. But so would the same amount of calorie-content served as plain milk, fruit, tea with honey, or any other of a long list of simple foods.

Laxatives. One common and often dangerous aid to reducing is the use of laxatives. They may be of the drug type or of the bulk-producing variety, or some other kind which stimulates the rate of passage of food mass through the body. They will, of course, reduce the amount of food absorbed from the intestinal tract, but they will also reduce the supply of the nutritional essentials, such as vitamins and minerals, and thus increase the chances for the development of deficiencies. "Slimming salts" and other preparations claiming to have the same effect fall into this category.

Thyroid aids. Another type of preparation that is sometimes sold to make reducing easier contains thyroid derivatives, or drugs that have a similar action. Their purpose is to speed up the body so that it runs at high speed and burns more fuel. Your doctor *may* find that a little thyroid will be beneficial in your particular case if you seem somewhat deficient in this hormone. But, he will prescribe it cautiously and in the minimum amount and check up on you frequently. And he generally will insist that you go on a strict diet as well. In other words, he will try to bring your metabolism up to normal so that your dieting efforts will have a fair chance of success. But you should not take such preparations by yourself. They are dangerous when not carefully controlled. They are stimulants and can be overdone easily.

Hunger depressants. Another variety of aids in reducing are those which are supposed to reduce your appetite. There are drugs in this class and some are too hazardous for use. However, none should be used indiscriminately. Your physician may allow you to take a limited amount of the safest one, but only if he can check up on the effects frequently. He may prescribe it for the first few weeks to help you through the transition period during which you are establishing a new eating habit. But it is just a crutch for temporary use and should not be used indefinitely. Other products which are supposed to decrease your hunger frequently contain a bit of sugar which is supposed to give your blood-sugar a boost for a time. They cost a great deal more than a small piece of candy, which would be just as effective. Remember that sugar and candy are to be avoided if you are to follow the diet prescribed in this book.

Miracle foods. There is a special class of foods, some of them very palatable and others very unappetizing, that are often considered to be "miracle foods." Of course, because they are "miracles" you have to pay special prices for them. They generally are harmless, but in addition they will not perform miracles. Wheat germ, brewers' yeast, blackstrap molasses, yogurt, vegetable juices (often mixed to order for each individual), and other similar products are common examples. Some are promoted as "cures" or as "glamour" foods, aids to beauty, long life and sex appeal.

Dried yeast *is* a food, of course, and it contains quite a number of vitamins, though not all. There is nothing in it, however, which you cannot get in your ordinary food.

Blackstrap molasses is not by any means as palatable as regular molasses. It is the residue from sugar refining—

what is left after all possible sugar has been removed—though it still contains sugars of various sorts. It is a waste product that finds use in cattle foods and in making alcohol. It contains some vitamins and iron and other minerals (partly as factory contaminants), but no extraordinary amounts, considering the relatively small amount you can use. Again, it supplies nothing you cannot get in regular food. But it *is* black and ill-tasting and mysterious and so you may be urged to buy it.

Sometimes, instead of blackstrap, you are urged to use raw sugar. In effect, this is cane sugar plus blackstrap. It is certainly no better—possibly a little worse. It has no special value even though it is urged as a "natural" food. There is no magic in the term "natural," so don't be misled.

Wheat germ *is* just what it says, the oily germ of the wheat which is separated out when the white flour is milled from the wheat berry. It contains about twenty-five per cent protein, ten per cent fat, some moisture and quite a bit of starch. There are generous amounts of the B vitamins and of Vitamin E. But ordinarily, you would use relatively small quantities and the contribution it would make *is* minor. It won't cure anything.

Yogurt is coagulated, fermented milk, generally made from whole milk or from milk plus cream. It has slightly less fat content than the milk from which it was made and generally is quite pleasant to eat. It contains lactic acid in place of some of the milk sugar, and so it resembles buttermilk. As in the case of milk, it is a valuable food—but it is no "miracle" food in itself.

* * * * *

Rx *for Reducing:*

1. **Food superstitions are as silly as any other superstition.**
2. **Most fad diets are unhealthy—and have no permanent value.**
3. **There are no "magic" foods to help you lose weight.**
4. **Stay away from fads. Eat a well-balanced high-protein diet for weight-reduction success.**

* * * * *

11: WHAT ABOUT EXERCISE?

Most people who want to reduce have two reasons for so doing: they want to look better and they want to feel better.

Exercise can be a big lift toward both of these goals—though you should be warned immediately that the most strenuous physical activity cannot work off enough pounds to count.

For instance, if you now weigh 160 pounds, you would have to swim steadily for eight hours to lose only one pound of fat. So you see that not only do you not have the time for this kind of reducing program, but you do not have the stamina to keep up such a program of weight loss. Furthermore, it would be a completely unfeasible way to lose weight.

However, just as important as the amount of fuel taken in by the body daily is how much it burns. Over a period of time, relatively small amounts of activity can greatly increase the body's burning of fuel. With a few slight changes in your routine activity you can accomplish almost as much as by limiting your food intake.

On the average, a 30-minute walk would burn up the equivalent of 150 calories. At this rate, it would take you twenty-three days to lose a pound—sixteen pounds in a year. Of course, the walk would naturally increase your appetite and you might eat a little bit more. Still, it is not unreasonable to say that a half-hour's daily walk could slice off five to ten pounds in a year.

Why is walking so much more effective than strenuous exercise? The answer is that the more work the human body does, the more energy it burns. In walking, the action is one which moves your entire body weight. You move the same amount of weight with exercise like a "push-up." But you use small muscles, muscles which are unsuited to the job, and you get tired before you do too many push-ups. Therefore, it is far more advisable to use the big leg muscles, muscles which are specifically designed to move larger weights longer distances.

But even using the bigger muscles, it still must be emphasized that the amount of energy you burn up in exercise is small compared to your average daily energy intake. For instance, just how much energy is spent in

ACTIVITY	CALORIES PER HOUR PER POUND OF BODY
Bowling	1.75
Cooking	.87
Dancing	2.10
Dishwashing	.87
Dressing, undressing	.71
Driving a car	.92
Eating	.67
Gardening	1.06
Golf	1.75
Ironing	.87
Knitting	.70
Laundry work	1.10
Lying in bed awake	.47
Painting furniture	1.20
Playing piano	1.16
Reading aloud	.63
Sewing	.67
Singing	.74
Sitting quietly	.60
Sleeping	.40
Standing relaxed	.63
Sweeping with:	
carpet sweeper	1.25
dust mop	1.02
vacuum cleaner	1.78
Swimming	3.02
Table tennis	2.65
Typewriting	.85
Walking	1.81
Washing floors	1.06
Writing	.67

Working out in the gym, surprisingly enough, doesn't burn up an excessive amount of energy. Light exercise will use up some 25 calories an hour; active exercise about 200 calories; and severe exercise only around 600 calories an hour.

However, while exercising is not going to burn off all those extra pounds you'd like to have disappear, a few gentle exercises followed faithfully can accomplish what might seem like a miraculous reduction of inches—though your poundage will remain the same unless you follow, with equal faithfulness, the diet in this book. Whether you are mountain-climbing or pruning the roses in your garden, you should, however, stop your exercise when you find yourself pleasantly tired. If you stop at that time, your quickened breathing and heart rate should return to normal within about ten minutes. And after an hour or two of complete rest there should be no weakness or fatigue whatsoever.

If muscular strain lingers into the next day and if your muscles ache, it's a good indication that your exercise has been too strenuous or too prolonged. In that case, you should hold off for a few days and then cut your exercising time to half and continue to cut or increase until you have established a normal pattern of activity so that there is no tired aftermath.

Just because you have allowed yourself to become one of the fifty-five million or more people who are overweight in this country, doesn't mean to say that you should allow yourself to join the even bigger group of Americans who are "sitters." It seems strange that even in the hustle-bustle of contemporary life, this country has developed into a nation of sitters. Because we sit so frequently, it turns out that much of our fatigue is simply a matter of not using our energy wisely. And once the muscles of the body slacken, the excess weight we are carrying becomes more and more obvious.

As the pounds creep up, the muscles of the waist and the abdomen stretch to make room for them. It becomes harder and harder to tighten them. It is far easier to let them expand than to hold these muscles in place.

As the waistline expands, it is easier to let the shoulders slump. That throws the stomach out still farther. Almost without noticing it, the overweight person adds years to his appearance by slipping into the habit of poor posture.

Therefore, aside from the increase of activity which is good for you—such as developing a hobby which will keep you standing for a period of time, rather than watching television—there are also some very simple exercises designed to improve posture and to tighten those stretched or relaxed muscles which have been carrying a burden of fat. This is not to say that you should not increase your ordinary activities. For instance, if you live ten blocks from your work, you ought to walk there and home again daily. In a month's time you'd probably burn up the equivalent of about one pound. But exercise is not going to strip away pounds quickly. Only diet can do that.

In any case, the posture exercises which are given below can be done anywhere and they should be done all day long. For example:

• When you sit down to breakfast, keep your posture in mind. Pull your head up as far as you can. Let your shoulders relax. Pull the stomach muscles in tightly. Try to keep your spine straight. Lean forward toward your plate from the hips instead of from the shoulders. You are going to find that automatically your belt feels looser around the waistline. Just because you are holding your head up and your stomach in, you will have achieved a look of greater youthfulness in place of the weary slump of age. You'll find that the double chin is a little less prominent. There will be less strain on the back of your neck. All in all, you are going to feel better—and feeling better, you will begin to look better.

• However, check your posture ten minutes later. Re member that it is not easy to reverse a long-established habit. You probably will find that you have let yourself slip back into that old slouching curve, so straighten up, and in so doing you will have done another posture exercize without anyone being the wiser. And you will notice the difference after a few days as the new habit replaces the old.

To assume glibly, however, that a few weeks of intensive exercise—or even fifteen minutes every day—will greatly change the contour of your body is naïve. All patrons of "slenderizing" salons or eager exercisers for short periods have discovered that failure to continue the exercising or treatments indefinitely means a return of the muscle laxness.

Yet, while exercise alone cannot solve the problem of overweight, here are a few exercises which will help you tighten up your muscles while this diet is melting away the pounds of fat.

• A good exercise for better posture is as follows:

1. Stand quite erect with your head up, your chin and rear tucked well in, your feet together and your shoulders, arms and hands relaxed.
2. Swing the arms crosswise (like a man trying to keep warm in cold weather) with the hands heavy and relaxed, the head still held up with chin tucked in. Inhale deeply as you do.
3. Rise on your toes, swinging your arms as high as you can, throwing your head back, and holding your breath, count up to ten.
4. Exhale, return to the first position. Repeat this exercise ten times.

• Another exercise that will help your posture:

1. Stand erect with your feet apart, your head held well up and your chin in, your shoulders relaxed. Tuck in your rear as well as you can. Inhale deeply.
2. Twist the upper body as far right as possible, swinging your right arm across. Follow your finger tips with your eyes.
3. Swing your right arm high up and far to the back. Keep

your elbow unbent. Twist the waist as you do so. Exhale.
4. Without returning to position one, start the reverse with position two. Repeat ten times.

• A good exercise to tighten up the abdominal muscles:

1. Lie on your back on a hard surface with your arms relaxed by your side.
2. Raise your legs, keeping your knees straight. If possible, get the legs straight up at a right angle to the body.
3. Lower your legs very, very slowly. As you do this you will feel all the muscles of your abdomen tighten. Rest a minute. Repeat ten times.

This last exercise is *not* an easy exercise to do at first, though it seems to be quite simple when you start it.

• Another good exercise for tightening the muscles of the abdomen:

1. Lie on your back on a hard surface with your hands behind your head.
2. Bend your right knee up to meet the left elbow; kick your toe up. Hold this position for a slow count of five.
3. Return to the first position. Reverse, sending the left knee up to meet the right elbow.

Remember that modern man no longer lives in a hunting-and-fishing-for-a-livelihood culture as did his ancestors of thousands of years ago. Therefore, he is not conditioned to vigorous exercise beyond the days of his youth. Vigorous exercise for most people is not healthful, nor is it desirable, because it may put persistent demands upon the heart and on the circulatory system beyond their normal capacity. You must avoid the damage that can result.

Don't allow the desire to create a "body beautiful" to destroy you before your allotted years. A healthy and attractive body is a worth-while goal, but a man of forty-five who wants to look like a college athlete is foolish indeed to train as seriously and as hard as the youth who is twenty-five years his junior and whom he is trying to emulate. There are exceptional people, of course, who have gone into their later years with the muscular bodies of youth. The difficulty is that most of us are less fortunately endowed. We would be placing a strain on the body far beyond its capacity if we were to follow their example.

But keep on exercising, gently, regularly—and happily. Steady, daily exercise will tone up your muscles, improve your circulation, and stimulate you into following your diet program.

* * * * *

Rx for Reducing:

1. **Exercise is not going to take off excessive pounds.**
2. **Exercise will tighten body muscles, make you look and feel better.**
3. **Never exercise excessively. Know your limitations.**

* * * * *

12: WEIGHT WATCHING

The battle against overweight is not one which can be easily won. While you can bring yourself down to the weight level you desire—perhaps with some ease, perhaps with some difficulty—the real difficulty, as you well know, lies after you have accomplished this aim.

Ninety per cent of the fifty-five million or more people in this country who are overweight today have tried at least once to lose weight. They have been successful in dropping a certain amount of weight, only to find that they were battling with the scale once again after they went off their diet.

The battle of the bulge is always followed by a period of cold war. In order to maintain your weight level, you have to become a "weight watcher." Just how long the cold war of weight watching goes on depends a great deal upon your reasons for losing excess weight. Some people have a strong motivation to lose weight temporarily. For example, young men and women will often lose their excess fat during a period of courting. In order to win a sweetheart, they want to look their best. A woman who feels that she is "losing her husband" may drop her excess weight and regain her more youthful slimness. A man who has been told by his physician that his years are numbered because of obesity may effectively lose weight.

Unfortunately, most often, when these people feel the danger has passed—or when they have attained the object which the weight reduction program was to bring for them—they soon begin to regain excess fattiness.

It is obvious, then, that most people who want to reduce do so because they wish to be more attractive, to feel better, or to live longer. Your doctor would probably reverse the order if we were asked to list the reasons why you should reduce. But let's look at these reasons in the other order, the order that you would probably put them into.

The desire to be attractive is the most frequent reason why most people want to reduce, whether they are men or women. You might call this desire a vanity, but it is based upon sound reasons. For instance, there is no doubt that in our present civilization the woman who is slender and the man who is trim are far more attractive to others than people who are overweight. The age has passed when "pleasant plumpness" was an indication of economic well-being.

There is a social handicap to obesity which we recognize today. The fat person is often left out of activities —or left till last. This is much more apparent with children than it is with adults, but the social exclusion does occur in adult life too.

There is often an occupational reason why slimness is desirable. Certainly most often on the stage, in modeling, and in other similar occupations—except for occasional specialized purposes—the slim person has a far better chance of winning a position.

The trim person also has an advantage economically over the heavier-set person. Better clothes can be purchased for the same amount of money or less. In buying food a considerable saving is possible. So this is a factor to be taken into consideration.

As far as feeling better is concerned, there is no doubt that being overweight brings with it personal discomforts that are unpleasant. There is the mechanical impediment which makes the overweight person clumsy and slow of movement. Breathing becomes less easy, a strain is placed on the body's circulation system and on the glandular system. When you are overweight your body works harder even when you are completely at rest.

And there are also psychological handicaps to being overweight. As was noted before, the overweight person often feels excluded from the activities of associates and therefore fails to develop the interests and social skills which are necessary for success and happiness in a normal adult life. The constant fear of embarrassment often makes him withdraw from normal activities. He may give up the sports that he enjoys and finally stay away from his friends. These changes in turn affect the personality. The obese person becomes more introspective and secluded. And often, if allowed to go for too long, the problem requires psychiatric aid.

The most important reason for reducing, as far as your doctor is concerned, is the need to be healthier. It is easy to shrug off the fact that you are a few pounds overweight—but it is dangerous. Life insurance companies have found, as you have already been told, that the mortality rates of overweight people are greater than those of persons of normal weight. Perhaps the best evidence to show that weight control pays has been compiled by the Metropolitan Life Insurance Company. Their specialists did a study of persons who were obese and who succeeded in losing sufficient weight so that they could qualify for insurance. This study showed that the death rate after weight reduction was substantially less than that for all obese persons studied. So, if you are too fat and fail to reduce, you are neglecting to take advantage of the most important single thing that you can do that will simultaneously make you look more attractive, feel better, and live longer.

How can you tell if you are overweight? Well, self-diagnosis is not too difficult. Actually, all you have to do is look in the mirror and you will see what your problem is. If there is some doubt, then your physician is best experienced to tell you whether you should reduce or not.

But, if you have any doubts as to what you should weigh ideally, why not look at the charts that follow. It is interesting to note that these new tables—compiled in

1959—differ considerably from the previous standards which had been based on two similar studies.

For example, the new average weights for women are quite consistently less than they were in the early studies; for men they tend to be higher than before.

Thus, at age twenty-five, the average weight for a woman is generally five to six pounds less than that shown in the last study. At the age of thirty-five, it is two to four pounds less. At the age of forty-five, it is generally two to three pounds less.

The decrease in average weights of women reflects chiefly their efforts to keep slim and, to some extent, the lighter weight of their clothing.

Among men, the increase in average weight, compared with earlier standards, varies appreciably with height. The increase is greatest for short men, amounting to five pounds or more at most ages. For men of average height, the increase is from two to four pounds, and for tall men, from one to three pounds. The larger differences are found under age forty-five.

Apparently the campaigns for weight control, to which women have clearly responded, have had much less impact on men.

Note that the tables of "desirable weights" include the wearing of "indoor clothing/' so that for men six pounds should be considered the average clothing weight, and for women four pounds. You will also note that the height shown is for a person wearing shoes. Therefore, one inch should be included for men's heels and two inches for women's heels.

These tables, then, give you an idea of approximately what you should weigh, though your doctor should be consulted so that he may take into consideration aspects which a table of "averages" cannot consider.

You, therefore, will attempt to get down to a weight that is "most desirable" for you. In so doing, following the high-protein diet, you will find that there will be large losses of weight during the first few weeks. These weight losses include large quantities of water. In later weeks,

AVERAGE WEIGHTS FOR MEN AND WOMEN
According to Height and Age

HEIGHT (IN SHOES)	WEIGHT IN POUNDS (IN INDOOR CLOTHING)					
	Ages 20-24	A es 25-29	Ages 30-39	Ages 40-49	Ages 50-59	Ages 60-69
	MEN					
5' 2"	128	134	137	140	142	139
3"	132	138	141	144	145	142
4"	136	141	145	148	149	146
5"	139	144	149	152	153	150
6"	142	148	153	156	157	154
7"	145	151	157	161	162	159
8"	149	155	161	165	166	163
9"	153	159	165	169	170	168
10"	157	163	170	174	175	173
11"	161	167	174	178	180	178
6' 0"	166	172	179	183	185	183
1"	170	177	183	187	189	188
1"	174	182	188	192	194	193
	WOMEN					
4' 10"	102	107	115	122	125	127
11"	105	110	117	124	127	129
5' 0"	108	113	120	127	130	131
1"	112	116	123	130	133	134
2"	115	119	126	133	136	137
3"	118	122	129	136	140	141
4"	121	125	132	140	144	145
5"	125	129	135	143	148	149
6"	129	133	139	147	152	153
7"	132	136	142	151	156	157
8"	136	140	146	155	160	161

*Average weights not determined because of insufficient data.
Source: *Build and Blood Pressure Study,* ***1959,*** **Society of** ***Actuaries.***

less and less weight will be lost as reflected on the scale; but sometimes the tape measure will show the results of the dieting while the scale will not.

You should be warned, however, that after achieving your desirable weight you can, by overeating, put on excess weight at about the same rate that it came off. You can put on as much as seven pounds in a week, and successive weeks will see four, then three, then two pounds added until you reach the level that you were when you started.

No matter how easy or how difficult it has been for you to achieve your normal weight in dieting, once having

—DESIRABLE WEIGHTS FOR MEN AND WOMEN
According to Height and Frame. Ages 25 and Over

HEIGHT (IN SHOES)	WEIGHT IN POUNDS (IN INDOOR CLOTHING)		
	SMALL FRAME	MEDIUM FRAME	LARGE FRAME
		MEN	
5' 2"	112-120	118-129	126-141
3"	115-123	121-133	129-144
4"	118-126	124-136	132-148
5"	121-129	127-139	135-152
6"	124-133	130-143	138-156
7"	128-137	134-147	142-161
8"	132-141	138-152	147-166
9"	136-145	142-156	151-170
10"	140-150	146-160	155-174
11"	144-154	150-165	159-179
6' 0"	148-158	154-170	164-184
	152-162	158-175	168-189
	156-167	162-180	173-194
		WOMEN	
4' 10"	92- 98	96-107	104-119
11"	94-101	98-110	106-122
5' 0"	96-104	101-113	109-125
[illegible]	99-107	104-116	112-128
4"	102-110	107-119	115-131
5"	105-113	110-122	118-134
6"	108-116	113-126	121-138
7"	111-119	116-130	125-142
8"	114-123	120-135	129-146
9"	118-127	124-139	133-150
10"	122-131	128-143	137-154
11"	126-135	132-147	141-158
	130-140	136-151	145-163

Note: Prepared by the Metropolitan Life Insurance Company. Derived primarily from data of the *Build* and *Blood Pressure Study*, 1959, Society of Actuaries.

reached your desirable weight you must constantly watch for many months—sometimes even years. The basic difference between sum people—people of normal weight— and overweight people is the amount of food intake and the unconscious control of this intake which permits the normal-weight person to maintain that normality unconsciously. Overweight people will not be satisfied with the amount of food involved in normal eating. They feel that this is quite inadequate. Thin, or normal-weight, people may overeat while in their overindulgence. Overweight people will at all times overeat or eat frequent meals interspersed with snacks. So, to stay at any specified weight, you must eat approximately the same amount all the time. If you reward yourself with a series of special treats every time you take off some weight, you are going to discover to your consternation that you have regained many pounds that you sweated hard to take off. This discovery can destroy your morale, especially if you are convinced that the struggle against overweight is a futile one.

But any number of pounds gained in one week of overindulgence after achieving normal weight can be removed in a short period of returning to the strict diet of this high-protein regime. It may mean that you'll have to forego the sensual pleasures of eating rich desserts—perhaps even permanently. In other words, to become a weight watcher, you must do *consciously* what people of normal weight do *unconsciously.*

The first step in watching your weight is knowing your exact weight status periodically. This is done by weighing yourself *once a week on the same scale, at the same time of day, wearing the same amount of clothes.* This will bring about the most exact correlation possible.

If you weigh yourself more than once a week you are going to become "scale-happy." Weighing more frequently than prescribed, you will never know your exact weight status. You see, there are many shifts in water balance in the body—such as during a woman's menstrual period, or a man's excessive perspiration—and these are reflected in a daily seesaw on the scale.

A person who has overindulged in food one day may gain a false sense of security the next if he should step on the scale and find that he has not gained any weight or, perhaps, has even lost a pound. This could encourage further excessive eating and within a week, to the weight watcher's dismay, many pounds will have been added.

Some people may find that after watching their food intake carefully over a period of weeks, their weight will have increased a few pounds. This is caused by temporarily retained water—and it can be a discouraging thing to discover.

But, by checking once a week, under the conditions outlined here, a better picture of your weight watching can be achieved. Record carefully your weight each week, as on the following chart.

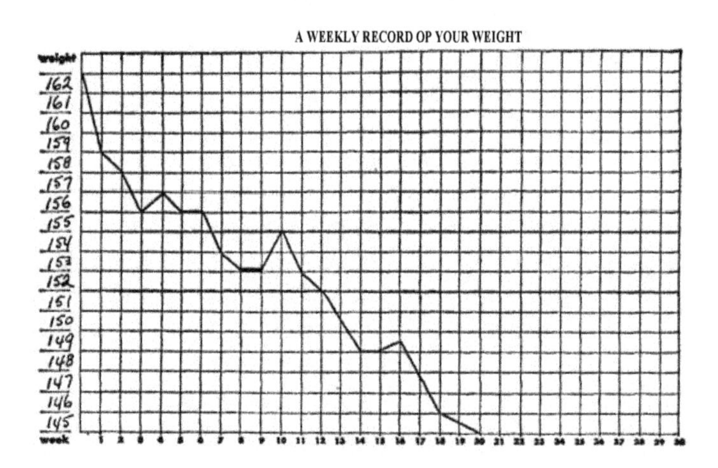

If a pound or two is added, then go into "training" immediately and remove it before something more serious is involved—and by more serious is meant greater weight increase, and more severe dieting restrictions.

It is possible that after achieving your normal weight you will continue to lose weight This is due usually to an overstrict diet regime after the normal weight has been attained. Weight watching will prevent further weight loss since, if you find that you continue to drop weight, you will be encouraged to eat a little bit more to hold your weight at the desirable level.

The first weeks of weight watching will indicate to you just what and how much food is necessary to maintain your normal weight. You may find that you must omit sweets, rich desserts, perhaps even tasty snacks permanently. You may find that restriction of food every second, or third, or fourth week is necessary. Best of all, you may find that you have permanently overcome your excessive desires for food and that you will be able to maintain your normal weight within a short time of having completed your diet regime—unconsciously.

Above all, the problem must never get out of hand. If you start to regain weight, if you cannot successfully cut back within the next few weeks, if you go over the top level—do *not wait* until you have regained all your weight before going to see your doctor to determine the reason why. If you wait too long, you will only involve yourself in a vicious seesaw of excessive obesity back to normal, back to excess—forever and ever. This is not a solution to the problem of overweight.

You can be assured that once you have lost your excessive weight through dieting and have carefully watched your weight, you will find that the problem of remaining at a desirable weight becomes less and less difficult as time goes on. The process soon becomes one of second nature and quite "painless."

* * * * *

Rx *for Reducing:*

1. **Successful reducing is a long-term project.**
2. **Getting down to the desired weight is not the end of the battle.**
3. **Once you've reached your goal, you must become a weight watcher.**
4. **Keep track of your own weight-reduction program on the charts on the following pages.**

* * * * *

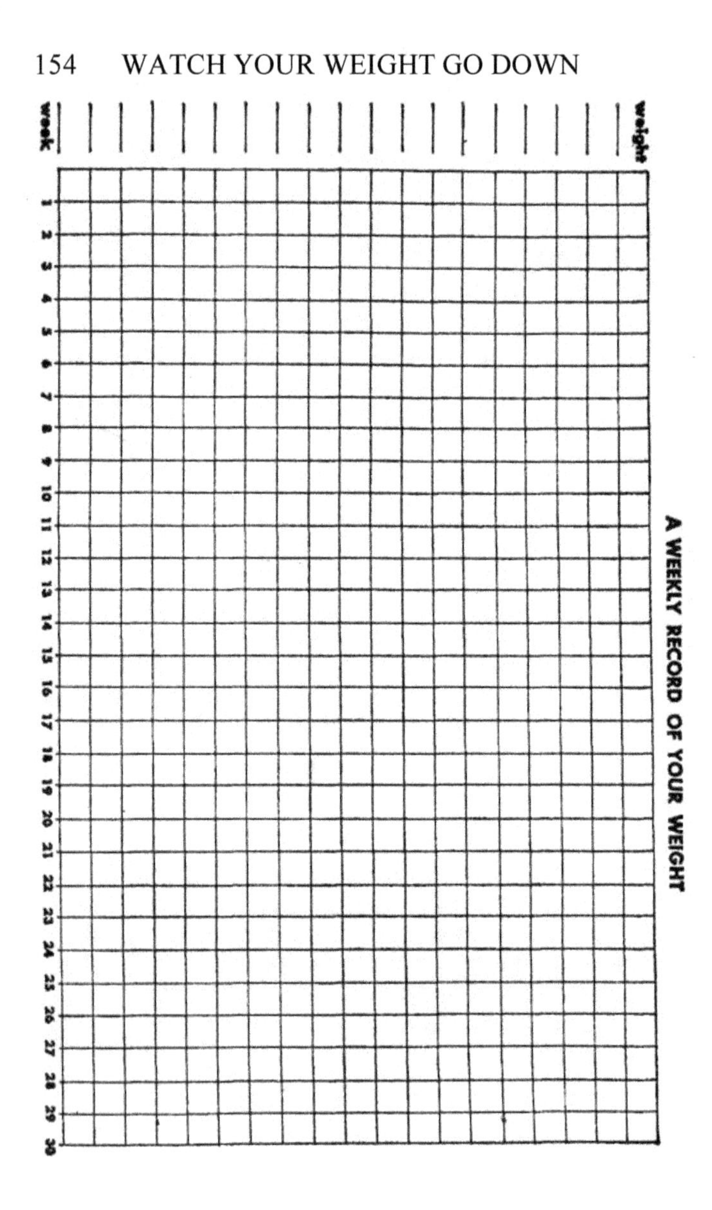
week
weight
1 2 3 4 5 6 7 8 9 10 11 12 13 14 15 16 17 18 19 20 21 22 23 24 25 26 27 28 29 30
A WEEKLY RECORD OF YOUR WEIGHT

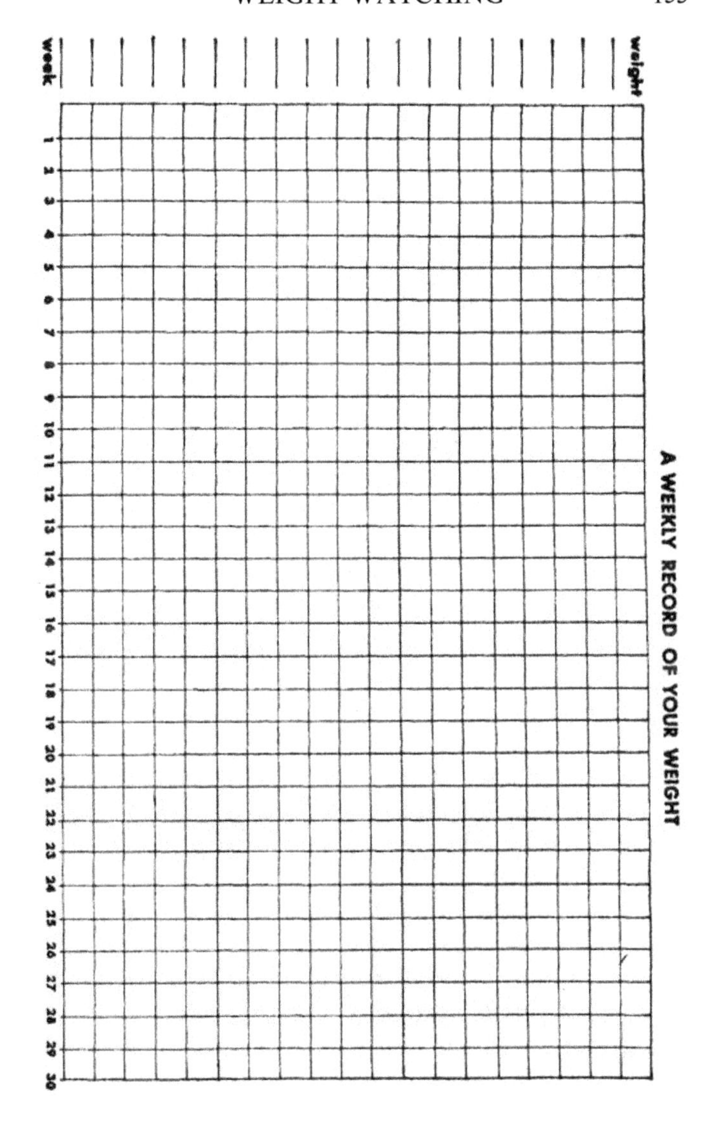
A WEEKLY RECORD OF YOUR WEIGHT
week
weight
1 2 3 4 5 6 7 8 9 10 11 12 13 14 15 16 17 18 19 20 21 22 23 24 25 26 27 28 29 30

13: DOCTOR ABOVE ALL

"If reducing is so easy and safe—just eating high-protein foods for a few weeks or a few months—why do all diet experts insist: *'See your doctor first?"*

Anybody—and everybody—who is overweight asks that question.

If you are ten pounds or so overweight, in good health, and are able to choose a sensible plan that will curtail your food intake enough to help you drop a pound or so a week, there is no real need for a medical checkup before starting.

But if you are fifteen pounds or more overweight, you're already in "bad shape." And, since your overweight problem is a serious one, you absolutely should consult your doctor.

This is not meant to scare you, but you must sensibly realize that the extra weight that you have been carrying around may have put too much strain on your heart, it may have contributed to a rise in. your blood pressure, or it may have speeded the progress of some other illness.

However, another problem is involved. The charts of desirable weights given in the previous chapter are all meant as "averages." But just what does "average" really mean? This is a question which in your particular case only your own physician can answer. This table of desirable weights can help you to compare yourself with others of the same sex, height and body build. But no single weight is correct for all people of the same height and sex. Some people have large, thick, bulky bones. Others have small, thin, delicate bones. Others are in between.

And so even with the charts being broken down for people of light frame, medium frame, or heavy frame, there are still certain discrepancies possible. Your doctor is best qualified, then, to determine just how overweight you are and how best you should reduce. He also should be kept informed of how successful—or unsuccessful (though, perhaps, we shouldn't mention this)—you may be on your diet program.

As was mentioned previously, as far as medicine is concerned, the need for better health is the most important reason why you should reduce. Insurance company studies have shown that the trim and the slender outlive the overweight. Yet, strangely enough, "obesity" is rarely if ever listed among the important causes of death. Except in very unusual cases, people are not listed as dying of overweight. The death from obesity is, however, found hidden in the figures for heart disease, for high blood pressure, for arteriosclerosis, for kidney disease, for diabetes, and even for cancer. Some of the increase in the death rate in all of these illnesses can definitely be accounted for by the increased obesity of the population in this country.

Diabetes is a double hazard for the person who is overweight. In the first place, the overweight person is more likely to develop diabetes, heart disease, high blood pressure and arteriosclerosis. In the second place, the person suffering from diabetes—even if he is of normal weight—is more likely to develop heart disease, high blood pressure and arteriosclerosis. Therefore, the obese person who develops diabetes is constantly threatened not just by one sword hanging over his head, but by two. The fact that people are different, and individual characteristics must be considered, makes it difficult to include in any calculation of how much you should weigh and how much weight you should lose. For that reason, your doctor is the person best qualified to determine this for you.

It is apparent, therefore, that you should not attempt to start, or to continue, a weight-reduction program without your physician's help. First, he is needed to help you understand why you are obese. There are very definite reasons why people overeat and only a doctor can help you understand this. Then, with his help, you can set about losing weight.

He will best help you fit your diet to your own needs, depending upon what vitamins and minerals you may require in larger amounts than necessary. He also may administer hunger-curbing medication should you require it.

But most important, he will help you pass through any of the trying periods that you may encounter while losing weight.

The most important thing, finally, is having your physician set a top level of weight for the future—once you have successfully lost the weight you wanted to lose. This level should never be exceeded. And the knowledge which you have acquired during your diet program should be utilized in keeping the balance and in reducing from this top level if it is ever reached.

If at any time you go over the top level, you should return to see your doctor again. Be warned that some people have one relapse, others may have two—some may even experience more. As a chronic dieter, you may have already experienced this. But, with the help of your doctor—and with a top-level figure set for you—it is likely that you can hold your desired weight for the rest of your life.

The problem must never get out of hand. If you start to regain weight, and if you cannot successfully cut back quickly and find yourself heading toward that top level, don't wait: see your doctor immediately.

° ° ° ° °

℞ for Reducing:

1. **See your doctor—immediately,**
2. **Follow his directions carefully.**
3. **Watch your weight constantly.**
4. **Good dieting—and good luck!**

° ° ° ° °

Also available from **www.sunvillagepublications.com**

Printed in the USA
CPSIA information can be obtained
at www.ICGtesting.com
LVHW020149050124
768088LV00042B/364

9 781438 288185